Information Practices and Knowledge in Health

Health Information Set

coordinated by
Céline Paganelli and Viviane Clavier

Volume 5

Information Practices and Knowledge in Health

Edited by
Céline Paganelli
Viviane Clavier

iSTE

WILEY

First published 2022 in Great Britain and the United States by ISTE Ltd and John Wiley & Sons, Inc.

ISTE Ltd
27-37 St George's Road
London SW19 4EU
UK

www.iste.co.uk

John Wiley & Sons, Inc.
111 River Street
Hoboken, NJ 07030
USA

www.wiley.com

Library of Congress Control Number: 2021949036

British Library Cataloguing-in-Publication Data
A CIP record for this book is available from the British Library
ISBN 978-1-78630-611-1

MIX
Paper from responsible sources
FSC® C013604

Contents

Introduction

This collective work, the fifth in the series *Health Information*, brings together eight contributions. It deals with health information and communication and highlights strong trends or "movements" in the field of health, which consist of many dimensions to be articulated in order to approach knowledge and information practices.

A first movement involves the growing visibility of health issues in the public space, an observation that dates back more than 20 years, as shown by issue 95 of the French journal *Réseaux* published in 1999 on "Science, the sick and the public space". This issue focused on patients and patient organizations that had acquired legitimacy in the public arena, thus contributing "to a new way of thinking about the relationship between science and the general public". This relationship, according to the authors [CAR 99, p. 10] "[was] less of an asymmetrical relationship of expertise and pedagogy, but rather [allowed for] the expression of scientific uncertainties, the sensitive involvement of caregivers or the claims of patients" [*ibid.*]. The media appear to be the key players in this publicizing of health issues. As early as the 1980s, television relayed appeals for donations or compassion for the Telethon [WAL 98]; the "media coverage" of AIDS or the story of contaminated blood provided the opportunity for "tests of strength" between the political, medical and judicial arenas, as well as for the internal tests of strength in the journalistic field on the "legitimate definition of information" [CHA 94]. Beyond the media space, society as a whole is, according to the health sociologist Pierre Aïach [AÏA 98] and the anthropologist, sociologist and doctor Didier Fassin [FAS 98], going through a process of

Introduction written by Viviane CLAVIER and Céline PAGANELLI.

medicalization that they consider to be a cultural transformation. Protecting ourselves and fighting against illness, detecting and preventing risks, staying healthy, etc. have become a constant preoccupation, elevated to the rank of a movement: healthism [CRA 82]. This "concept", which emerged during the 1970s and 1980s, refers in particular to the Foucauldian thesis according to which individuals come to adhere to social norms of their own free will, even though they are not forced to do so. According to Hélène Pouliquin, healthism is recognized as "an elevation of health to the rank of virtues and morality, accompanied by an individual duty to maintain health by making individuals responsible, even if at the same time the state disengages from acting on the social, political and economic determinants of health" [POU 15, p. 26]. Promotional discourse and consumer practices are part of this trend, as Benoit Lafon shows with another neologism he calls "sanitarization" to indicate, once again, a double movement perceptible in the "consumer press" [LAF 20]: on the one hand, the consideration of public health issues in entire sectors of the economy, which, beyond agriculture, concern various industrial activities such as agri-food, chemistry or electronics. On the other hand, the increasing consumerization of healthcare is manifested in the growth of medical devices, "reimbursed" by complementary health insurance organizations in competition with each other (glasses, hearing aids, etc.), or by the ever-expanding dynamics of the drug industry towards other areas to be medicalized (well-being, smoking, etc.). This same tendency can be found in food, as attested by a third neologism coined in 2009 by the food sociologist Jean-Pierre Poulain [POU 09]. Thus, the "nutritionalization" of food refers to the management of food in the context of a specific pathology by a health professional specialized in nutrition. It also refers to the massive diffusion of nutritional knowledge in health education, prevention campaigns and the media.

A second movement is linked to the reconfiguration of public service in hospitals and health institutions, and to the evolution of forms of management that aim to reconcile a number of "complex and manifold challenges" [GRE 19]. The context is marked by cost reduction, rationalization of professional activities, inequalities in social and territorial access to healthcare, the aging of the population and the increase in chronic diseases. Nevertheless, the healthcare on offer must always be improved in order to ensure patient satisfaction. Thus, the development and modernization of information systems for public and private institutions is part of a policy to improve administrative, logistical and care processes [VIG 18]. Among these systems, the Electronic Patient Record (EPR), although not recently introduced, is emblematic of the evolution of hospital practice and is the result of both the need for traceability of actions and decisions associated with patient care and the need for coordination between specialists in charge of patients [MOR 18]. The deployment of EPRs in healthcare institutions has made it possible, thanks to surveys of health professionals, to specify expectations in terms of information (relevance, availability, reliability, security, real-time updating, etc.); integration of heterogeneous data or definition of knowledge meta-models to promote semantic interoperability between terminologies [CAB 17]. Thus, the organization of knowledge and the information practices of professionals are at the heart of reflections on the design or uses of these health information systems.

A third movement concerns patients taking responsibility for their own health. The notion of "health democracy", which appeared in the political field and was used for the first time by the public authorities in the early 2000s, covers citizen participation in health policies [ARV 16] and was officially included in the functioning of the French healthcare system by the law of March 4, 2002, relating to the rights of patients and the quality of the healthcare system [LET 09]. The role of users, sick or healthy, of their relatives or of patients' associations is thus recognized when it comes to health issues [LEF 18]. Combined with the development of liberal economic policies in the health sector, "health democracy" seems to have led to an increase in control regulations [DEM 14], the emergence of the patient-client, thus promoting a consumerist approach to health [BAT 08], or even the instrumentation of users, as denounced by certain associations [LEF 18]. Nevertheless, this evolution has been accompanied by the development of systems for accessing health information, allowing patients to consult information or share their experiences; associations to publish documentation on the Internet or to structure communities for exchange and sharing. Thus, the user takes control of health issues by discussing the issues with medical professionals, supporting a loved one, or dealing with a diagnosis or the evolution of an illness, in order to find practical information or verify scientific data. According to a 2016 IPSOS survey, more than 68% of French people have already consulted the Internet to find medical information, whether it be via collaborative sites, general sites dedicated to health or institutional sites [CAS 16]. The medical profession, friends and family, and the media are also sources of information that are sought after. A real knowledge base is then developed, "lay expertise" that emerges among patients. This expertise is based on lived experience and considers that people, through their personal experience of an illness, acquire specific knowledge on the issue. It is recognized by users of discussion forums or social health networks, who consult these systems to solicit testimonies from patients or former patients [PAG 11]. It also consists of real medical and scientific expertise. Indeed, patients, often through associations that bring them together, invest in the scientific field, appropriating specialized knowledge in order to make their voices heard and participate in the governance of the healthcare system [AKR 12]. The development of this form of expertise is facilitated by access to specialized information resources via the Internet. Although the notion of the expert patient is now well established, it in fact conceals a very contrasting reality. First, because it reduces the patient's identity to their illness [GRI 20], and second because it does not take the diversity of situations into account. The availability of information resources and easy access to digital devices do not guarantee the appropriation of knowledge. Studies on health literacy, which the WHO defines as "the personal characteristics and social resources needed by individuals and communities to access, understand, evaluate and use information and services to make health decisions"[1], show the diversity of the necessary skills (informational, media-related, medical, etc.) and the constant need to update these skills [MAR 17]. The implementation of mediation actions or systems, relying in particular on documentary mediators [TET 20], therefore appears essential. Finally, using therapeutic education to improve health literacy would contribute to supporting patients in the appropriation of knowledge related to disease and, more generally, to reducing inequalities in access to care [MAR 17].

A fourth movement concerns the development of digital devices in the service of health, a sector that is experiencing various kinds of initiatives and constant growth. Estimated at 2.7 billion euros in 2014, the e-health market is dominated by innovative start-ups specializing in telemedicine[2]. These include technologies that enhance health knowledge[3], such as those that take

1 See: https://www.who.int/healthpromotion/conferences/9gchp/health-literacy/en/.

2 Pôle interministériel de prospective et d'anticipation des mutations économiques, Prospective. E-santé : faire émerger l'offre française en répondant aux besoins présents et futurs des acteurs de santé, Synthesis, Edition Marine Automne, Nicole Merle-Lamoot, 2016. Available at: https://www.entreprises.gouv.fr/files/files/directions_services/etudes-et-statistiques/prospective/Numerique/2016-02-Pipame-e-sante-synthese.pdf.

3 Picard R., Serveille H., Vial A., Technologies et connaissances en santé, Public report, December 2013. Available at: https://www.vie-publique.fr/sites/default/files/rapport/pdf/273726.pdf.

medical investigation further or reduce surgical risks, and that are used to rationalize and control the production of care. We should also mention other applications, such as medical imaging, training devices (for surgical procedures, for example) for learning, individual training and simulation, or even connected objects, which are "composed of sensors designed to transmit information via a mobile application or a web service" [BEY 17]. More focused on data than on knowledge, applications intended for self-quantification aim to collect, measure and compare various variables in order to improve well-being, and maintain or improve health status [CAM 16]. The incursion of new players in this very lucrative market – only 32% are from health sectors such as medical industries, insurance, hospitals and physicians according to [STA 18] – reveals different concerns and interests. Adrian Staii identifies the economic models that regulate this market, as well as the actors who continue to invest [STA 18, pp. 217–218]: 28% are pure players who develop activities in the field of mobile health; 23% are actors outside the health sector (IT and digital industries) and 11% are non-market organizations (associations, NGOs, etc.). Thus, connected health may prove to be one of the most blatant indicators of the loss of autonomy of traditional health players in a market now dominated by digital players, who are legitimate in prescribing standards in the well-being market [*ibid*, p. 220].

The book is organized in eight chapters. The first two chapters focus on the field of health information and its evolution in the context of Open Access, and aim to characterize health information from the point of view of its uses and associated practices.

Focusing on scientific information, Chérifa Boukacem-Zeghmouri and Hans Dillaerts report on the regulations induced by Open Access, which has developed significantly in recent years in all scientific fields. However, its adhesion and legitimization are particularly significant in the health field. Plan S, which encourages the publication of research results from public grants in journals or platforms that respect the founding principles of open science, has strongly changed the model of scientific communication in the health field, since scientific research in this field is essentially based on calls for proposals. Moreover, the introduction of new publication models and new actors (Gold publishers, megajournal publishers, preprint servers or predatory publishers), the positioning of historical actors who produce medical information within STM publishing, new practices in terms of publication, dissemination strategies, valorization and mediatization of knowledge all take new criteria into account, such as the general public or reputation on social networks. The researchers situate their reflections in the current context (2020–2021), that of the Covid-19 pandemic, and show that the health crisis highlights the importance and complexity of the issues of publication and dissemination of validated scientific information. Finally, they raise the question of the future of scientific journals, which are not equipped and organized to disseminate research results in real time, to the benefit of preprint platforms, which receive researchers' publications as they progress.

Between opportunities and risks, this first chapter provides elements for understanding the changes underway for all of the actors in scientific and technical information (STI) within the health field.

Chapter 2, written by Céline Paganelli and Viviane Clavier, offers a methodological contribution aimed at characterizing health information by studying the information practices of health professionals and the organization of knowledge. Thus, returning to the documentary perspective, which approaches treatments according to their uses, health information is presented as strongly linked to contexts of use and practices. Based on the results of field studies and corpus analyses conducted in the field of health for more than 10 years, the researchers propose to rediscuss the contours of health information and put forward elements that allow it to be characterized.

The following chapters aim to understand how actors, whether professionals or citizens, inform themselves, how questions of evaluation of the information collected arise in a sector where the reliability of information is crucial and how health literacy can be considered in digital environments or in relation to specific populations.

In Chapter 3, Eloria Vigouroux-Zugasti, Olivier Le Deuff and Amar Lakel propose a reflection on health literacy in digital information environments. A concept that appeared in the early 1970s, in parallel with that of information literacy, health literacy goes far beyond the framework of informational or media skills and is becoming, particularly with the evolution of health information provision, a public health issue and challenge. In this chapter, the authors question digital health literacy based on the results of field studies: an automatic analysis of web corpora containing health information, as well as a longitudinal qualitative survey on online health practices among a retired population, using digital technologies. The results highlight several trends: the dominant place of the press and health professionals in the health information available on the Internet and the very minor voice of patients. They also show the need for users to develop skills that will enable them to be more comfortable when searching for health-related information.

Digital health literacy covers a variety of skills (informational, digital, media-related, health-related, etc.) that are difficult to evaluate because they require regular updating. The acquisition of these skills is a matter of training, as well as of mediation mechanisms, the reinforcement of which, by relying on health professionals and mediation actors, seems to be a necessity. However, beyond the aspects relating to training and mediation, the chapter clearly shows that public health policies must take up the issue of digital health literacy insofar as it constitutes a major challenge for therapeutic education.

Chapter 4 is written by two Finnish researchers from the discipline of Library and Information Science: Kristina Eriksson-Backa and Stefan Ek. These researchers are interested in the health information behavior of older people – their interests, the sources they use, the reliability criteria given to information sources, etc. – and propose methods to assess their level of health information literacy and medical knowledge. The notion of information behavior is borrowed from Tom Wilson[4] and the empirical results were obtained through a survey sent by mail to 1,000 Finns aged 65–79 years. A total of 281 responses were completed and returned by a majority of women (57%), almost half of whom were aged 65–69 years and had an intermediate level of education (44%).

Health information literacy refers to skills related to information literacy, the ability to recognize a need for health information, evaluate and use this information in daily life to make good health decisions, use information ethically and legally and so on. Most studies relate literacy levels to the age of individuals and their ability to read and understand health-related materials. There are many questionnaires designed to measure health literacy. Recent studies have shown an association between low health

4 "The totality of human behavior in relation to sources and channels of information, including both active and passive information seeking, and information use (Wilson 2000, p. 49)".

literacy, poorer health status and increased use of hospital care. Conversely, better health literacy could be an important factor in the adherence to prescribed treatments. One of the contributions of the authors' research is to highlight a link between the mastery of health information and the perception of patients when communicating with health professionals.

The next two chapters address issues of prevention in health information and communication.

Thus, Chapter 5, written by Cécile Loriato, proposes to study information on HIV/AIDS prevention from the perspective of the categorization of audiences and the hierarchization of actors in the journalistic narrative. Since the end of the 2000s, new biomedical prevention tools have been proposed, including Pre-Exposure Prophylaxis (PrEP), which is used for HIV prevention by HIV-negative people and is mainly recommended for men who have sex with men (MSM). The publicization of PrEP has sparked controversy around homosexuality, HIV and risk because it calls into question years of prevention based on condom use as the only known and effective tool. In this context, the researcher studies the construction of the journalistic narrative on biomedical prevention. She examines, on the one hand, the use of epidemiological categories to designate people at risk and, on the other hand, the hierarchical structure of the journalists' account of prevention, in order to understand how this contributes to the total or partial exclusion of certain actors, actions, events or problems. The results show that information on biomedical HIV/AIDS prevention is constructed mainly from the point of view of scientific experts and mobilizes mostly stigmatizing epidemiological categories.

Through this analysis, the researcher contributes to the characterization of media information on prevention. Journalistic discourse on prevention gives a predominant place to scientific expertise. Indeed, the views on the biomedical prevention strategies of doctors and researchers specializing in HIV/AIDS are predominant, while the views of associations and users of these means of prevention are hardly heard. This reliance on experts can be explained in part by the fact that some journalists are not very close to the field of HIV/AIDS, which leads them to give preference to information from the communication services of medical or health institutions which have institutional legitimacy.

Chapter 6, written by Aude Chauviat, also addresses prevention information, as disseminated in prevention campaigns. She situates her argument in a particular historical context, that of the National Prohibition campaign in the United States (1919–1933). Based on content analysis of a corpus of more than 200 pamphlets published by the Anti-Saloon League, as well as a sub-corpus of 82 pamphlets mobilizing health information and created for the most part by the Scientific Federation for Temperance, she shows how health information was mobilized in the prevention campaign against alcohol to the detriment of a moralistic approach to the issue. She postulates that this shift to a preventive discourse based on health information is similar to a form of medicalization of alcohol-related problems and marks an evolution towards a *medical* approach to health, made possible by the acquisition of legitimacy of scientific discourse among the general public.

This contribution proposes a contextualized approach to health information. While anti-alcohol prevention campaigns rely more on specialized information, it is because for one, the discipline and international scientific community specializing in the question of alcohol is becoming institutionalized and more structured, and in parallel, popularized scientific discourses aimed at the American population are emerging.

Finally, the last two chapters address health information from the perspective of knowledge organization and representation.

Chapter 7 presents the testimony of Dr. Four, a private practitioner specializing in ophthalmology, on the role played by professional journals in updating medical knowledge for the practice of his profession. This testimony is very instructive for research on the information practices of health professionals and on the organization of knowledge. Indeed, Dr. Four has been regularly monitoring professional journals for many years (24 years). This activity is integrated in his weekly schedule, since he devotes 15 hours of his time to reading journal articles and processing the bibliography. To practice general ophthalmology, it appears that updating knowledge is essential in many areas of ophthalmology, some of which evolve more rapidly than others. This interview is an opportunity to indicate the fields in which information is more scarce or, on the contrary, those in which it is abundant but, paradoxically, little summarized. These different cases require greater efforts in terms of information research, whether to find rare sources, summarize numerous publications or validate sources. Thus, information practices are fully integrated into medical practice and the information obtained constitutes a strong point in support of the diagnosis, assessment, treatment protocol, referral of patients to other specialists or justification of therapeutic decisions from a medical–legal point of view. This knowledge, built up and developed over time, is recorded in a document base, the organization and structuring of which are designed *by* and *for* Dr. Four's medical practice, and is presented in detail in the chapter.

Chapter 8 by Marcin Trzmielewski and Claudio Gnoli is a joint contribution on the evolution of medical knowledge organization systems (MKOS) from the 1960s to 2019. These researchers are both specialists in knowledge organization, a field of activity and research that focuses on the design, study and critique of methods for organizing and representing knowledge, most often implemented in libraries and documentation centers. Based on a review of 71 articles published in information and documentation sciences, this chapter describes the main families of MKOSs, the most represented being ontologies, thesauri and classifications. The increase in the number of publications on this subject since 1960 reflects both the vitality of the field of knowledge organization, particularly in the United States and Canada, and the strong dynamics of scientific production in medicine, and the generalization of health information on the medical Internet. Beyond the study of the structuring of concepts, the relationships between concepts, and the evolution of the various fields of medicine, this chapter presents a complete analysis of the "general public" or specialized users of MKOS. The study also focuses on the document processing process and questions the testing of specialized classifications organized by discipline in the face of interdisciplinarity. It appears that this field is still dominated by concerns related to the representation of knowledge, for which classifications also raise socio-cultural or political questions.

References

[AÏA 98] AÏACH P., DELANOË D. (eds), *L'ère de la médicalisation. Ecce homo sanitas*, Anthropos, Paris, 1998.

[AKR 12] AKRICH M., RABEHARISOA V., "L'expertise profane dans les associations de patients, un outil de démocratie sanitaire", *Santé Publique*, vol. 1, no. 1, pp. 69–74, 2012.

[ARV 16] ARVEILLER J., TIZON, P., "Avant-propos : démocratie sanitaire, qu'est-ce à dire ?", *Pratiques en santé mentale*, vol. 2, no. 2, p. 2, 2016.

[BAT 08] BATIFOULIER P., DOMIN, J., GADREAU M., "Mutation du patient et construction d'un marché de la santé. L'expérience française", *Revue française de socio-économie*, vol. 1, no. 1, pp. 27–46, 2008.

[BEY 17] BEYALA L., *Connected Objects in Health: Risks, Uses and Perspectives*, ISTE Press, London and Elsevier Ltd, Oxford, 2017.

[CAB 17] CABOT C., Recherche d'information clinomique dans le Dossier Patient Informatisé : modélisation, implantation et évaluation, Computer Science PhD thesis, Normandy University, 2017.

[CAM 16] CAMBON, L., "Objets connectés, mobiles, communicants en prévention : dépasser l'outil, penser l'intervention…", *Santé Publique*, vol. 28, no. 1, pp. 5–6, 2016.

[CAR 99] CARDON D., HEURTIN J.-P., "Présentation", *Réseaux*, vol. 17, no. 95, pp. 9–12, 1999.

[CAS 16] CASSAN M., Rapport d'étude : usages et attentes des Français à l'égard du digital en matière d'information sur leur santé, Report, Ipsos MSD, May 2016.

[CHA 94] CHAMPAGNE P., MARCHETTI D., "L'information médicale. sous contrainte", *Actes de la recherche en sciences sociales, L'emprise du journalisme*, vols 101–102, 1994.

[CRA 82] CRAWFORD R., "Healthism and the medicalization of everyday life", *International Journal of Health Services*, vol. 10, no. 3, pp. 365–388, 1982.

[DEM 14] DEMAILLY L., "Variations de la 'démocratie sanitaire' et politique publique de santé mentale en France", *Sociologies*, 2014.

[DOD 99] DODIER N., "L'espace public de la recherche médicale, autour de l'affaire de la ciclosporine", *Réseaux, Sciences, malades et espace public*, vol. 17, no. 95, pp. 107–154, 1999.

[FAS 98] FASSIN D., "Avant-propos : les politiques d'une médicalisation", in AÏACH P., DELANOË D. (eds), *L'ère de la médicalisation. Ecce homo sanitas*, Éditions Economica, Paris, 1998.

[GRE 19] GRENIER C., BERTRAND P., HERVE H., "Quels renouvellements conceptuels pour soutenir l'innovation dans le champ de la santé ? Un regard par les arrangements organisés, les politiques publiques et les capacités entrepreneuriales", *Innovations*, vol. 60, no. 3, pp. 5–14, 2019.

[GRI 20] GRIMALDI A., "Patient expert ou patient ressource ?", *Médecine des maladies métaboliques*, vol. 14, no. 6, pp. 545–548, 2020.

[LAF 20] LAFON B., "'Our health in danger': The extension of sanitization through media coverage of health alerts", in PAILLIART I. (ed.), *New Territories in Health*, ISTE Ltd, London and John Wiley & Sons, New York, 2020.

[LEF 18] LEFEUVRE K., OLLIVIER R., *La démocratie en santé en question(s)*, Presses de l'EHESP, Rennes, 2018.

[LET 09] LETOURMY A., NAÏDITCH M., "L'émergence de la démocratie sanitaire en France", *Santé, Société et Solidarité*, no. 2, pp. 5–22, 2009.

[MAR 17] MARGAT A., GAGNAYRE R., LOMBRAIL P. *et al.*, "Interventions en littératie en santé et éducation thérapeutique : une revue de la littérature", *Santé Publique*, vol. 6, no. 6, pp. 811–820, 2017.

[MOR 18] MORQUIN D., OLOGEANU-TADDEI R., "The Electronic Medical Record: Standardization issues and personalization of information for health professionals", in PAGANELLI C. (ed.), *Confidence and Legitimacy in Health Information and Communication*, ISTE Ltd, London and John Wiley & Sons, New York 2018.

[PAG 11] PAGANELLI C., CLAVIER V., "Le forum de discussion : une ressource informationnelle hybride entre information grand public et information spécialisée", in YASRI-LABRIQUE E. (ed.), *Les forums de discussion : agoras du XXIe siècle ? Théories, enjeux et pratiques discursives*, L'Harmattan, Paris, 2011.

[POL 15] POLIQUIN H., "Analyse critique et dimensionnelle du concept de santéisation", *Aporia*, vol. 7, no. 1, pp. 17–29, 2015.

[POU 09] POULAIN J.-P., *Sociologie de l'obésité*, Presses Universitaires de France, Paris, 2009.

[STA 18] STAII A., "Connected health: Between common aspirations and specific interests", in PAGANELLI C. (ed.), *Confidence and Legitimacy in Health Information and Communication*, ISTE Ltd, London and John Wiley & Sons, New York, 2018.

[TET 20] TETE C., "Les documentalistes comme partenaire dans la prise en charge des patients en fin de vie", *Médecine Palliative*, vol. 19, no. 1, pp. 3–6, 2020.

[VIG 18] VIGNERON B., "Chapitre 6. Le management opérationnel des systèmes d'information", in LOUAZEL M. (ed.), *Le management en santé : gestion et conduite des organisations de santé*, Presses de l'EHESP, Rennes, 2018.

[WAL 98] WALTER J., *Le Téléthon : Scène – Intérêts – Éthique*, L'Harmattan, Paris, 1998.

1

Open Access as a Regulation of Scientific Information in Health

1.1. Introduction

At the time of writing (April 2020), a pandemic was spreading across the world, claiming the lives of hundreds of thousands of people. Medical research is at the front line in finding a suitable vaccine against Covid-19. More broadly, however, all research in the field of health contributes to possible responses to the disease, and to its wider management (development of Covid-19 tests by virologists, large-scale implementation of tests by veterinary researchers, etc.). A recent preprint [COL 20] has shown – through a scientometric approach – the breadth of the health fields mobilized around Covid-19 research [BRA 20]. Scientific journals and preprint servers have seen the number of submitted articles soar [HOR 20].

The pandemic has led to a policy dynamic, driven by funding agencies, that requires all publishers to make all articles dealing with Covid-19 research immediately available. The Wellcome Trust[1] was the first, followed by the call for data from the US White House[2]. The International Coalition of Library Consortia (ICOLC) also called upon organizations, scientific

---

Chapter written by Chérifa BOUKACEM-ZEGHMOURI and Hans DILLAERTS.

1 "Sharing research data and findings relevant to the novel coronavirus (COVID-19) outbreak": https://wellcome.ac.uk/coronavirus-covid-19/open-data.

2 "Call to Action to the Tech Community on New Machine Readable COVID-19 Dataset": https://www.whitehouse.gov/briefings-statements/call-action-tech-community-new-machine-readable-covid-19-dataset/.

communities and publishers to ensure immediate Open Access to scientific publications and research data[3].

The biomedical fields are well and truly present, but they are poorly represented in the acronym STM, which conveys a definition that is more representative of the economic and industrial actors (notably, the scientific publishers and their powerful STM association) than of the research fields and their communities. Hence, throughout this chapter, we have chosen to use the term "Health", with the first letter capitalized. It is more unifying and closer to the French definition of the biomedical sector discussed in this chapter. However, one of the characteristics that allows these fields to constitute a sector is the role and the place that Open Access to scientific information has represented in the development of their content since the 1990s.

This chapter aims to give an account of the regulations generated by Open Access and which today link the field of scientific communication with the field of Health. We will address the way these regulations affect the publication and dissemination models of digital scientific information. The aim is to provide an account of the nature of these regulations and the issues they raise regarding scientific certification, regulations and issues that are crucial in Health.

1.2. The political anchoring of Open Access in the field of Health

The public and societal importance of the field of Health led to it being one of the first fields to switch from paper to digital journals in the late 1990s. As Momen [MOM 03] points out,

> Health is perhaps the area of most intense demand for greater access to scientific and technical information, partly because failure to obtain it can be literally fatal.

When experimenting with the first electronic journal projects, publishers always included biomedical titles in their corpus [WIL 94]. This is not insignificant because, perhaps more than in any other field, the high financial stakes that govern health research place the scientific publication system at

3 "Statement on the Global COVID-19 Pandemic and Its Impact on Library Services and Resources": https://icolc.net/statement/statement-global-covid-19-pandemic-and-its-impact-library-services-and-resources.

the heart of the mechanisms of permanent competition. With the deployment of the Web in the early 2000s and the rules of its attention economy [CIT 13], Open Access to scientific information has even become a value [RIF 05] around which public policies have turned to reach out to researchers, with recommendations and requirements. As a reminder, according to the Budapest Declaration of 2002[4], Open Access is defined by two complementary channels. The Green Road – or self-archiving – characterized by archiving and online publication of a copy of articles in an open archive, possibly with an embargo period. The Gold Road, characterized by a publication with free, open and immediate access in an Open Access journal.

In the early 2000s, the United States developed and ultimately mobilized open access to health information. The famous National Institute of Health (NIH), a government agency directly dependent on the Department of Health, was a major player and in 2004 adopted a policy in favor of Open Access, supported by the PubMed Central database, essential for researchers in the field. Researchers were invited to archive a copy of their published article on PubMed Central no later than 6 months after its publication [ZER 04]. Four years later, in 2004, this incentive became mandatory, with a 12-month period. As Moorhead *et al.* [MOO 15] point out,

> As a policy leader and major funder of health research, the NIH has, through its Public Access Policy, placed health literature at the forefront of the move toward open access to research.

In fact, unlike fields such as high-energy physics, where communities have mobilized around the principles of Open Access and have set up Open Repositories [GIN 97], in Health, it is public policies that, in light of the public health issue, have placed Open Access at the heart of thinking behind models for the dissemination of validated scientific information. Peter Suber [SUB 08] emphasizes the unprecedented nature of the approach, which was initiated by United States legislation and implemented by funding agencies.

The pioneering approach of the NIH spread like wildfire and saw other repository mandates and similar policies develop around the world [PIW 18]; the National Science Foundation (NSF), the European Commission (EC), the Wellcome Trust and, even more recently, the

4 "Budapest Open Access Initiative": https://www.budapestopenaccessinitiative.org/read.

Bill & Melinda Gates Foundation [BUT 17] to name just a few of the funding agencies embracing the same Open Access policy principles for the researchers they fund.

A few months after the launch of the compulsory mandate of the NIH, the European Commission launched its own Open Access pilot project as part of the 7th European Framework Program[5]. This pilot project – which accounted for 20% of the total budget of the framework program – was allocated to seven research themes considered priorities in the context of the construction of the European Research Area, including Health. Researchers receiving European funds had to:

> Strive to provide open access to these articles within six or twelve months – set according to the field of research – after publication.

In the framework of Horizon 2020 where Health is a major strategic focus, Open Access to research results (publications and data) has become the new standard for scientific and technical information (STI) dissemination. Commission Recommendation (EU) 2018/790 of April 25, 2018 on access to and preservation of scientific information[6] asked member states to:

> Define and implement clear policies [...] for the dissemination of and open access to scientific publications resulting from publicly funded research.

The success of these aligned policies can be measured by the 5.9 million articles published worldwide. And as has been pointed out [TEN 16], the cumulative number of Open Access articles has increased far more than the number of non-full-text articles. Meanwhile, during the pandemic, downloads of articles indexed with the keyword "Covid-19" were in the millions [KOU 20] and were not generated by researchers alone. Members of the general population, wanting to be informed, accessed these articles using search engines.

---

5 "European Commission launches online pilot project to facilitate access to scientific articles on EU-funded research", European Commission press release (2008): https://ec.europa.eu/commission/presscorner/detail/fr/IP_08_1262.

6 "Commission Recommendation (EU) 2018/790 of 25 April 2018 on access to and preservation of scientific information": https://op.europa.eu/fr/publication-detail/-/publication/2ea66d3f-649a-11e8-ab9c-01aa75ed71a1.

1.3. Gold Open Access or the "Author-Pays" principle

Building on the policies that promote Open Access in Health and aiming to speed up access to information, BioMed Central (BMC) is a new player making its way onto the digital scientific publishing scene. In the early 2000s, it developed an offer of digital journals *ex nihilo* that published articles in Open Access. To do so, it relies on the Gold model (characterized by publications with free, open and immediate access for the author and the reader) and is included in the Berlin Declaration for Open Access. However, BMC "adapted" the original Gold model by asking the author (or their institution) to pay to publish, in order to find the resources to develop its platform and model. BMC thus introduced the article processing charge (APC), a publication fee that allows the publisher to finance the publication, which is in fact distributed under a Creative Commons License. This is an ingenious reversal of the subscription business model: payment is no longer made by the reader (or their library) but by the author (their institution or research project). The Gold Author-Pays model is born [TEN 16]. This transformation led to more precise terminology: the original Gold model was renamed the Gold Diamond model, to emphasize the difference. The need for immediate access to validated scientific information in the field of health has thus given rise to a business model that is now at the center of structural changes in scientific communication in Health.

In 2006, PloS (Public Library of Science), another Gold publisher, launched a range of journals under the Gold Author-Pays model using the same arguments [BAR 06],

> The potential benefits of such a change are vast. No longer will physicians have to base their practice on half-truths. Instead, everyone from patients to policy-makers can read for themselves the evidence on which crucial science and health policy decisions are made.

The success of Gold Author-Pays publishing was still in its early stages: BMC, which became the spearhead of the young anthill of Gold publishers in the field of Health, was acquired by Springer in 2008[7]. It thus proved its success and, at the same time, transferred its expertise to the number 2 in

7 "Springer to acquire BioMed Central Group", Springer press release (2008): https://www.springer.com/about+springer/media/pressreleases?SGWID=0-11002-6-805003-0.

scientific publishing. The SpringerNature group, which is today one of the major players in Open Access publishing,[8] capitalized on BMC's experience to develop its Gold journal offer and the Read & Publish[9] agreements to libraries and their institutions, as part of the "*transformative agreements*". It is intended to replace subscriptions to digital journal packages [OLS 20]. BMC's success story will be emulated by others, and this is how scientific publishing will welcome a new category of Gold publishers, with technological and economic expertise in the distribution of digital scientific information on Health.

However, the Gold Author-Pays model raises questions about APC funding. The inequality of researchers in terms of access to information is now transformed into inequality in relation to publishing [BOU 17c]. Even though the Open Access model advocates access to knowledge, publishers are seizing on it to make it a new source of funding for their economic model. This is confirmed by the fact that the policies of some countries with a tradition of research and publication, such as the UK, officially opt for Gold Open Access as a strategic policy for the dissemination of scientific publications and provide, as a corollary, grants that allow researchers to publish according to this Gold Author-Pays model [FIN 12]. Since then, the major historical groups of scientific publishing (Elsevier, Wiley, Springer, etc.), the Gold publishers (PloS, Copernicus, etc.) and the international databases (Scopus, Web of Science, Dimensions, etc.) have set out to work on the development of Gold journals and their indexing, and thus on their economic valuation.

The evolution of the scientific publishing sector in the field of Health, subject to policy requirements, has therefore encouraged the development of the Gold model, and particularly the Gold Author-Pays model. According to Piwowar *et al.* [PIW 18], Health fields (biomedical research, clinical medicine, health, biology) are the disciplines in which there are the most Open Access scientific articles. Many studies have highlighted the strong growth of Open Access journals in the biomedical sciences since the

---

8 STMPublishing (2020), "Springer Nature achieves new milestone in 2019 in publishing over 100,000 OA articles in one year, and is the largest OA publisher of primary research": https://www.stm-publishing.com/springer-nature-achieves-new-milestone-in-2019-in-publishing-over-100000-oa-articles-in-one-year-and-is-the-largest-oa-publisher-of-primary-research/.

9 "Read and Publish (Springer Compact) agreements": https://www.springer.com/gp/open-access/springer-open-choice/springer-compact/springer-contact-intro-/293018.

beginning of 2002 [BOU 17a]. For example, more than half of all published scientific articles are available in Open Access [PIW 18].

1.4. Predatory journals: the "price" of change

It is no longer possible to address Gold Open Access journals without the phenomenon of predatory journals [COB 18] being included in the debate. Designated as such by Jeffrey Beall's blog in 2012[10], predatory journals are part of the collateral damages resulting from the development of the Gold Author-Pays model. One of the major characteristics of the phenomenon is that these journals use the Gold Open Access model and solicit authors, asking them to submit articles that they publish on their site without any form of peer review. A predatory journal therefore simply cashes in on the APCs without adding any value – notably scientific (no peer review process) – or any editorial improvement to the manuscript. A recent article proposes a definition that is already a collectively accepted reference [GRU 19]:

> Predatory journals and publishers are entities that prioritize self-interest at the expense of scholarship and are characterized by false or misleading information, deviation from best editorial and publication practices, a lack of transparency, and/or the use of aggressive and indiscriminate solicitation practices.

The phenomenon, which was in its infancy in the early 2010s, has grown to an estimated 14,000 predatory journals[11]. They present a significant risk to the field of Health, more so than for any other field. The largest number of these "predatory offers" occurs in the field of Health [GRU 19]. Predatory journals within this field therefore add to the phenomenon of Fake Science [HOP 19] and contribute to the complex, nebulous and dubious world of bio-medical information on the Web. The community of health researchers is aware of the danger this represents and is mobilized to denounce it. The extensive literature on the subject, published in medical journals, bears witness to this [CAM 18].

---

10 "Beall's List: Potential, possible, or probable predatory scholarly open-access publishers": https://web.archive.org/web/20170112125427/https://scholarlyoa.com/publishers/.

11 Cabells International (2020). The Blacklist: https://www2-cabells-com.ezproxy.haifa.ac.il/about-predatory.

The most recent research on this new subject reveals that Open Access might be a mechanism by which external actors penetrate and gain a foothold in the scientific communication and publishing scene, and build legitimacy [COB 18; SIL 20]. They also highlight the capacity of Open Access to implement hybridization and thus increase the complexity of publication models, whose intelligibility contributes to the reliability of the information they convey [DIL 18].

1.5. The place and weight of funding agencies in the OA environment

Changes in scientific communication in the digital era and in Open Access have broadened the spectrum of stakeholders, thereby influencing the definition and scope of dissemination models. In Health, in particular, research funding agencies, which historically did not intervene in the definition of scientific information dissemination models, have joined the Open Access equation. Today, they carry a lot of weight, as do publishers, libraries and learned societies.

Funding agencies very quickly began paying attention to the rise of Open Access. The most important among them introduced recommendations in the 2000s; these were directed at funded researchers and stipulated that published results be made available in Open Access. For example, the Wellcome Trust, one of the best endowed agencies in the world, adopted a mandate for Open Access in October 2006 [WAL 06]. Researchers receiving funding were required to provide open access to the results of their research.

The recommendations, clearly in favor of access to and visibility of funded research, are part of a management logic that finds its functional counterpart in the identification of these results in international databases, such as PubMed, Web of Science (WOS) or Scopus. Publications (mainly articles) are therefore not only "units of knowledge" but also "accounting units" [GIN 18] on which economic mechanisms of return on investment (ROI) are based.

Researchers previously invited to publish their results in "excellence" journals adapted and learned to archive their articles in thematic or institutional repositories [FRY 11]. Publishers have also adapted their Open Access policies to allow authors to deposit different versions of their papers (preprint, postprint author, postprint editor) according to specific deadlines and conditions (e.g. Sherpa Romeo).

Open Access journals (based on the Gold Author-Pays model) that appeared in the field of Health were a rapid success with authors who met all the expectations of research assessment: publication of articles in an Open Access journal, guaranteeing the visibility of research results. The authors had no difficulty justifying payment of the APCs from their research budget to the funding agencies. The latter even made it clear in their calls for projects that publication expenses (including APCs) were eligible for funding.

In Health, where research is based on a dual dynamic of collaboration and international competition, the financial stakes are huge; Open Access is therefore at the heart of the strategies of funding agencies (public and private) to give visibility and recognition to the research they have funded. This strategy has therefore been integrative of the various channels of Open Access, while at the same time driving momentum towards new avenues.

1.6. Plan S, a "radicalization"[12] of Open Access in Health?

The role of funding agencies took on a particular dimension in September 2018 with the announcement of an international coalition of 13 agencies (including the *Agence Nationale pour la Recherche* – ANR, French National Research Agency) encouraging researchers to publish in Gold Open Access journals [ELS 18]. The launch of Plan S at the initiative of Science Europe – the association of research funding agencies of the member states – reflects the desire to reinvent the methods of scientific communication by prioritizing the publication of research results from public grants in scientific journals or platforms that respect a number of founding principles of Open Science[13].

The announcement was like a thunderclap in the sky of scientific publishing because it took the funding agencies out of the prescriptive role of Open Access dissemination to put them (with the very first draft of

12 Adjective borrowed from [ELS 18].

13 Source: cOAlition S, Making full and immediate Open Access a reality: https://www.coalition-s.org/wp-content/uploads/271118_cOAlitionS_Guidance.pdf.

Plan S) in the position of favoring one model (Gold and *a fortiori* Gold Author-Pays) over another (Green representing the Open Archives)

Decried, criticized or welcomed, Plan S is still being debated, two years after its announcement. It is not unanimously supported in all fields[14], for obvious reasons of funding mechanisms. However, not all fields rely on project-based research. Furthermore, it has recently been shown [LAR 18] that the funding criterion does not necessarily enter into the researcher's publication strategy: as soon as authors have the funds to pay for the APC, they prefer to publish in Gold Open Access journals. *A fortiori*, in Health, where the overwhelming majority of research is project-based, Plan S acts as a catalyst for a restructuring of the scientific communication model around the imperative of access.

The Plan S coalition has been joined by new members including the influential Wellcome Trust; this new policy applies from January 1, 2021[15]. Today, the coalition includes powerful and important public and private funding agencies, who now feel they have a say in how science is communicated and disseminated. In addition to national funding agencies, four private international funding agencies are also associated with Plan S. All four fund Health research: the Bill & Melinda Gates Foundation, World Health Organization, TDR, the Special Programme for Research and Training in Tropical Diseases, and ASAP (Aligning Science Across Parkinson's). Some of them do not hesitate in their support for experimental Open Access publishing projects that challenge the journal model in favor of the stand-alone article. In doing so, they contribute to a shared reflection on the sustainability or obsolescence of the journal model, with regard to the new requirements for media visibility of validated science [BOU 19a]. As the STM 2018 report [JOH 18] points out,

> Strictly speaking, open access is a property of the written outputs of a specific research project, rather than the publication that hosts such outputs. In other words, it is a property of an article rather than a journal, or of an individual monograph rather than a book series.

14 In the Humanities and Social Sciences, Plan S has been heavily criticized by communities that do not have the financial means to consider APC payments, even though these are far below the APC rate within Health: https://plansinhss.home.blog/.

15 Source: https://wellcome.ac.uk/news/wellcome-updates-open-access-policy-align-coalition-s.

But above all, what justifies fears about Plan S is the fact that, while it builds its justification on a rhetoric that borrows from the arguments of Open Access, it undeniably introduces market regulation mechanisms to steer the publication strategy of researchers. The management logic of research is integrated into it, and there is a risk that it does not sufficiently take into account the ethical risks that are their corollary.

1.7. Scientific information in Health: a world of no journals?

Preprint servers are now an integral part of the scientific communication ecosystem in Health [PEN 20]. They have appeared over the last 10 years and illustrate the dynamism and importance of this method of dissemination [TEN 18; PEN 20]. Supported and set up by researcher communities, particularly early-career researchers [NIC 19], these servers represent a new generation of thematic open archives that offer an advanced digital service around preprint.

The most famous are those grouped around the extension "Xiv" of the eponymous archive arXiv, which founded the Green Road of Open Access. Using this extension in their names as a distinguishing mark conveying values, the new preprint servers claim both a filiation and a specific identity [BOU 19b]. In addition to openness, a discursive marker of Open Access, their rhetoric also includes "acceleration" as a strong justification, which indirectly points to the unsuitability of journals to ensure publication within a sufficiently short period of time.

A representative example is bioRxiv.org, a preprint server launched in 2013 in the biomedical sciences. It now lists over 77,800 preprints and has grown exponentially since its inception[16]. More recently, in 2019, we saw the birth of medRxiv.org in the medical field. A recent study lists up to 44 preprint servers within the Health domain [KIR 20], attached to funding agencies, public research actors or even scientific communities. These servers have attracted the attention of funding agencies that take into account the articles posted to these platforms. This started with the NIH, which promotes the appropriation of bioRxiv and medRxiv by researchers [ABD 19]. The researchers themselves are on board and have shown unwavering enthusiasm since 2018 [ELS 19].

16 Source: https://rxivist.org/stats.

The interest of these platforms is that they have become spaces for exploration and experimentation. In a "Bibliodiversity" approach[17] aiming at diversifying the models of Open Access publication, preprint servers, equipped with proofreading functionalities (comments, open peer review, etc.), alternative metrics or Altmetrics (mentions in social networks, downloads, etc.) make it possible to consider new ways to shake up the "old" model of the scientific journal [BOU 17b]. Thus, it is not insignificant to see that one of the preprints on the treatment of Covid-19 by Professor Raoult, available through biorXiv, has been downloaded more than 600,000 times, giving its author the opportunity to talk about it in TV interviews, presenting this extensive consultation as a guarantee of relevance and scientific credibility.

In the same vein, preprint repositories are becoming observation sites for scientific publishers and for research funding agencies. Publishers identify articles that are "buzzing" and offer to submit them to their journals [PEN 20]. Or, conversely, articles submitted to journals are automatically uploaded and shared on preprint servers like PLoS journals that allow authors to directly deposit the submitted article on bioRxiv[18]. On the other hand, bibliometric databases, such as Dimensions, or the academic search engine Google Scholar, are starting to take these servers into account and are adapting their algorithms for counting article citations. Authors have in fact integrated the issues of visibility and impact into their strategies for valorizing their research through, among other things, preprint repositories such as bioRxiv [SEV 19]. Fu and Hughey [FU 19] have thus demonstrated that:

> Articles with a preprint [on bioRxiv] had, on average, a 49% higher Altmetric Attention Score and 36% more citations than articles without a preprint.

These preprint platforms are now attracting the attention of private funding agencies, which are showing increasing interest in new Open Access publishing models. In 2016, the Wellcome Trust decided to launch its own publishing platform: Wellcome Open Research [BUT 16].

---

17 Jussieu Call for Open Science and Bibliodiversity: https://jussieucall.org/.

18 "Power to the Preprint", The official PLOS Blog: https://theplosblog.plos.org/2018/05/power-to-the-preprint/.

> The idea behind Wellcome Open Research is to allow Wellcome grant recipients to publish their findings more quickly and to create a model that, according to the charity, other funders might adopt in future.

In addition, as of 2019, the NIH has established an Open Access publishing pilot in partnership with the publisher PLoS to cover the cost of publishing in PLoS[19] journals. The Bill & Melinda Gates Foundation has also set up a publishing platform, Gates Open Research, which, like that of the Wellcome Trust, aims "to accelerate the publication of articles and data from research funded by the charity".

These initiatives are part of the processes of platformization [MIR 18] currently operating in scientific communication and are based on the following elements: rapid dissemination and publication of research results (concerns and aspirations found in both the open archives movement and mega-journals), upstream funding of articles by research programs, open science methods (release and publication of research data; open peer review) and, finally, new forms of mediatization of scientific and technical information as previously specified. This platformization questions the place of traditional health journals insofar as they are no longer the unique way to disseminate research results.

Large national funding agencies and private research funding organizations are thus helping to (re)shape the scientific communication landscape with Open Access publishing platforms [ROS 18].

1.8. Conclusion

The field of Health is currently in a period in which the issues of publishing and disseminating validated scientific information have perhaps never been so crucial. The Covid-19 health crisis has demonstrated its importance as well as its complexity.

Open Access is largely part of this complexity, through a political injunction, which is carried by not only public research organizations but

19 Source: https://www.nihlibrary.nih.gov/plos.

also public and private funding agencies. The latter now intervene directly in the determination of dissemination models and in the legitimization of new models of scientific communication. What we are witnessing in Health is the redefinition of a model of scientific publication that is more directly linked to the organization and financing of research.

The Covid-19 pandemic has suggested possible channels of scientific communication that are becoming independent of the journal model in favor of near-real-time dissemination of articles on preprint servers. Research “in the making” requires an acceleration for which journals are not ready; it therefore turns to platforms that play the role of intermediary device, facilitating and accelerating dissemination.

The question of scientific validation, and thus of peer review, remains the ultimate lock allowing journals to keep their central place in the scientific communication system, and giving information its guarantees of integrity and credibility. But here again, recent events, such as the LancetGate affair, illustrate[20,21] to what extent the walls of scientific publishing are becoming increasingly cracked in the field of Health. The rise of predatory journals and the phenomenon of Fake Science in recent years question more broadly the status of scientific soundness, integrity and credibility of scientific and medical information that circulates on the Web and that affects both scientific communities and the general public.

Open Access to scientific information in Health has not only improved the dissemination and circulation of scientific information but also reshuffled the cards of the game between stakeholders, allowing some of them, notably the research communities and funding agencies, to take a more important part in the definition of models that are better able to take care of the imperatives of contemporary scientific practice.

20 Morin H., Cabut S., Larousserie D., Santi P., Herzberg N., “Coronavirus : le “Lancetgate” révèle des failles de l’édition scientifique”, *Le Monde*, June 15, 2012, https://www.lemonde.fr/sciences/article/2020/06/15/covid-19-le-lancetgate-revele-des-failles-de-l-edition-scientifique_6042946_1650684.html.

21 The LancetGate has created a controversy around the Peer Review system, as it is practiced today, without necessarily taking into account the data collected and giving rise to the analyses.

1.9. References

[ABD 19] ABDILL R.J., BLEKHMAN R., "Tracking the popularity and outcomes of all bioRxiv preprints", *eLife*, vol. 8, p. e45133, available at: https://doi.org/10.7554/eLife.45133, 2019.

[BAR 06] BARBOUR V., CHINNOCK P., COHEN B. *et al.*, "The impact of open access upon public health", *Bulletin of World Health Organization*, vol. 84, no. 5, pp. 339–400, 2006.

[BOU 17a] BOUDRY C., DURAND-BARTHEZ M., "Publications en libre accès en biologie–médecine: Historique et état des lieux en 2016", *Ethics, Medicine and Public Health*, vol. 3, no. 1, pp. 169–181, 2017.

[BOU 17b] BOUKACEM-ZEGHMOURI C., CORDONNIER S., SPANO W. *et al.*, "'Publicize or Perish': Nouvelles formes de circulation et d'évaluation des savoirs scientifiques sur le Web", *Les Cahiers de La SFSIC*, no. 14, pp. 279–284, 2017.

[BOU 17c] BOUKACEM-ZEGHMOURI, C., NICHOLAS, D., "La vie en gold: Enjeux et risques pour les chercheurs", *I2D – Information, données & documents*, vol. 54, no. 1, pp. 10–11, 2017.

[BOU 19a] BOUKACEM-ZEGHMOURI C., RODRIGUEZ-BRAVO B., "L'information scientifique à l'épreuve de sa médiatisation", *Les Enjeux de l'information et de la communication*, available at: https://lesenjeux.univ-grenoble-alpes.fr/2019/articles-revue/dossier-2019-l-information-scientifique-a-l-epreuve-de-sa-mediatisation/, 2019.

[BOU 19b] BOUKACEM-ZEGHMOURI C., BERTHAUD, C., "The stakes of repositories and bibliodiversity: Community approaches", *ELPUB 2019 23rd Edition of the International Conference on Electronic Publishing*, Marseille, France, June 2019.

[BRA 20] BRAINARD J., "Scientists are drowning in COVID-19 papers. Can new tools keep them afloat?", *Science*, available at: https://doi.org/10.1126/science.abc7839, 13 May 2020.

[BUT 17] BUTLER D., "Gates foundation announces open-access publishing venture", *Nature*, vol. 543, no. 7647, p. 599, 2017.

[CAM 18] CAMACHO M., RECKLEY, L.K., "Predatory journals: Enough is enough", *Laryngoscope*, vol. 128, no. 7, p. 1510, available at: https://doi.org/10.1002/lary.27178, 2018.

[CIT 13] CITTON Y., "Économie de l'attention et nouvelles exploitations numériques", *Multitudes*, vol. 3, no. 3, pp. 163–175, 2013.

[COB 18] COBEY K.D., LALU M.M., SKIDMORE B. *et al.*, "What is a predatory journal? A scoping review", *F1000Research*, available at: https://doi.org/10.12688/f1000research.15256.2, 2018.

[COL 20] COLAVIZZA G., COSTAS R., TRAAG V.A. *et al.*, "A scientometric overview of CORD-19", *bioRxiv*, available at: https://doi.org/10.1101/2020.04.20.046144, 2020.

[DIL 18] DILLAERTS H., "Paysage de l'Open Access: Complexité et hybridation des modèles de publication et de communication scientifique", *URFIST de Rennes*, available at: https://drive.google.com/file/d/13_T-bvDwO-UiH86o7lqHJtAcDaStEv2K/view, 2018.

[ELS 18] ELSE H., "Radical open-access plan could spell end to journal subscriptions", *Nature*, available at: https://doi.org/10.1038/d41586-018-06178-7, 4 September 2018.

[ELS 19] ELSE H., "How to bring preprints to the charged field of medicine", *Nature*, available at: https://doi.org/10.1038/d41586-019-01806-2, 6 June 2019.

[FIN 12] FINCH J., Accessibility, sustainability, excellence: How to expand access to research publications, Report, Working Group on Expanding Access to Published Research Findings, RIN, 2012.

[FRY 11] FRY J., PROBETS S., CREASER C. *et al.*, PEER behavioural research: Authors and users vis-à-vis journals and repositories. Final report, available at: http://www.peerproject.eu/fileadmin/media/reports/Final_revision_-_behavioural_baseline_report_-_20_01_10.pdf, 2011.

[FU 19] FU D.U., HUGHEY J.J., "Releasing a preprint is associated with more attention and citations for the peer-reviewed article", *eLife*, available at: https://doi.org/10.7554/eLife.52646, 2019.

[GIN 97] GINSPARG P., "Winners and losers in the global research village", *The Serials Librarian*, vol. 1997, nos 3–4, pp. 83–95, 1997.

[GIN 18] GINGRAS Y., "Les transformations de la production du savoir: De l'unité de connaissance à l'unité comptable", *Zilsel*, vol. 2, no. 2, pp. 139–152, available at: https://doi.org/10.3917/zil.004.0139, 2018.

[GRU 19] GRUDNIEWICZ A., MOHER D., COBEY K.D. *et al.*, "Predatory journals: No definition, no defence", *Nature*, vol. 576, pp. 210–212, 2019.

[HOP 19] HOPF H., KRIEF A., MEHTA G. *et al.*, "Fake science and the knowledge crisis: Ignorance can be fatal", *Royal Society Open Science*, vol. 6, no. 5, available at: https://doi.org/10.1098/rsos.190161, 2019.

[HOR 20] HORBACH S., "Pandemic publishing: Medical journals drastically speed up their publication process for Covid-19", *bioRxiv*, available at: https://doi.org/10.1101/2020.04.18.045963, 2020.

[JOH 18] JOHNSON R., WATKINSON A., MABE M., *The STM Report. An Overview of Scientific and Scholarly Publishing: 1968–2018*, International Association of Scientific, Technical and Medical Publishers, The Hague, 2018.

[KIR 20] KIRKHAM J.J., PENFOLD N., MURPHY F. *et al.*, "A systematic examination of preprint platforms for use in the medical and biomedical sciences setting", *bioRxiv*, available at: https://doi.org/10.1101/2020.04.27.063578, 2020.

[KOU 20] KOUSHA K., THELWALL M., "COVID-19 publications: Database coverage, citations, readers, tweets, news, Facebook walls, Reddit posts", *Arxiv* (pre-print), available at: https://arxiv.org/ftp/arxiv/papers/2004/2004.10400.pdf, 2020.

[LAR 18] LARIVIÈRE V., SUGIMOTO C.R., "Do authors comply when funders enforce open access to research?", *Nature*, available at: https://doi.org/10.1038/d41586-018-07101-w, 24 October 2018.

[MIR 18] MIROWSKI P., "The future(s) of open science", *Social Studies of Science*, vol. 48, no. 2, pp. 171–203, available at: https://doi.org/10.1177/0306312718772086, 2018.

[MOM 03] MOMEN M., "Equitable access to scientific and technical information for health", *Bulletin of the World Health Organization*, vol. 81, no. 10, p. 700, available at: https://www.ncbi.nlm.nih.gov/pmc/articles/PMC2572335/, 2003.

[MOO 15] MOORHEAD L.L., HOZMEYER C., MAGGIO L.A. *et al.*, "In an age of open access to research policies: Physician and public health NGO staff research use and policy awareness", *PLOS ONE*, vol. 10, no. 7, available at: https://doi.org/10.1371/journal.pone.0129708, 2015.

[NIC 19] NICHOLAS D., WATKINSON A., BOUKACEM-ZEGHMOURI C. *et al.*, "So, are early career researchers the harbingers of change?", *Learned Publishing*, vol. 32, no. 3, pp. 237–247, available at: https://doi.org/10.1002/leap.1232, 2019.

[OLS 20] OLSSON L., LINDELÖW C.H., ÖSTERLUND L. *et al.*, "Cancelling with the world's largest scholarly publisher: Lessons from the Swedish experience of having no access to Elsevier", *Insights*, 33, available at: http://doi.org/10.1629/uksg.507, 2020.

[PEN 20] PENFOLD N.C., POLKA J.K., "Technical and social issues influencing the adoption of preprints in the life sciences", *PLOS Genetics*, vol. 16, no. 4, p. e1008565, available at: https://doi.org/10.1371/journal.pgen.1008565, 2020.

[PIW 18] PIWOWAR H., PRIEM J., LARIVIÈRE V. *et al.*, "The state of OA: A large-scale analysis of the prevalence and impact of Open Access", *Peer J*, vol. 6, p. e4375, available at: https://doi.org/10.7717/peerj.4375, 2018.

[RIF 05] RIFKIN J., *L'Age de l'accès, la nouvelle culture du capitalisme*, La Découverte, Paris, 2005.

[ROS 18] ROSS-HELLAUER T., SCHMIDT B., KRAMER B., "Are funder open access platforms a good idea?", *SAGE Open*, pp. 1–16, available at: https://doi.org/10.1177/2158244018816717, 2018.

[SEV 19] SEVER R., ROEDER T., HINDLE S. *et al.*, "BioRxiv: The preprint server for biology", *bioRxiv*, available at: https://doi.org/10.1101/833400, 2019.

[SIL 20] SILER K., "Demarcating spectrums of predatory publishing: Economic and institutional sources of academic legitimacy", *SocArXiv*, available at: https://doi.org/10.31235/OSF.IO/6R274, 2020.

[SUB 08] SUBER P., "An open access mandate for the national institutes of health", *Open medicine: A Peer-Reviewed, Independent, Open-Access Journal*, vol. 2, no. 2, pp. 14–16, available at: https://www.ncbi.nlm.nih.gov/pmc/articles/PMC3090178/, 2008.

[TEN 16] TENNANT P.T., WALDNER F., JACQUES D.C. *et al.*, "The academic, economic and societal impacts of Open Access: An evidence-based review", *F1000Research*, vol. 5, p. 632, available at: https://doi.org/10.12688/f1000research.8460.3, 2016.

[TEN 18] TENNANT J., BAUIN S., JAMES S. *et al.*, "The evolving preprint landscape: Introductory report for the knowledge exchange working group on preprints", *MetaArXiv*, available at: 10.31222/osf.io/796tu, 2018.

[WAL 06] WALPORT M., KILEY R., "Open access, UK PubMed central and the wellcome trust", *Journal of The Royal Society of Medecine*, vol. 99, pp. 438–439, 2006.

[WIL 94] WILLIS K., ALEXANDER K., GOSLING W. *et al.*, "TULIP – The University Licensing Program: Experiences at the University of Michigan", *Serials Review*, vol. 20, no. 3, pp. 39–47, 1994.

[ZER 04] ZERHOUNI E.A., "NIH public access policy", *Science*, vol. 306, no. 5703, p. 1895, 2004.

2

Knowledge Organization and Studies of Information Practices: Contribution to the Characterization of Health Information

2.1. Introduction

The field of health information and communication has undergone profound changes since the mid-1990s. Health information is protean and can be divided into four categories:

– specialized medical information, produced by specialists for specialists (scientific articles, for example);

– professional information regulated by public authorities, which supports the activity of health professionals: patient files, health insurance cards;

– public health information produced by institutions within the framework of prevention campaigns or surveys of the population, or, in the field of food, nutritional information regulated by health safety agencies;

– and finally, information for the general public: although the popularization of health information in the media is not new, its development has taken on a new dimension with the Internet, the digitization of resources and the development of free access. In particular, social media allows specialized information to be made available to the public.

Chapter written by Viviane CLAVIER and Céline PAGANELLI.

Health information, whatever form it takes, is produced and edited by many actors: publishers, health institutions, professional or patient associations, professionals in the field or even ordinary citizens, most often those affected by a particular pathology. Private actors are an important part of this, since the health information sector is a financially attractive market. The Web in the first instance and then the "social Web" have enabled the development of numerous health info-communication systems. Examples include specialized databases reserved for scientists and professionals that have become accessible online and can be consulted by the general public; the sites of health institutions or general public information sites giving access to forums or social networks specialized in specific pathologies.

Changes in the way health information is produced and disseminated are accompanying a number of changes in the medical field. On the one hand, shared medical records provide healthcare professionals with information that can be useful for coordinating patient care. Introduced in 2004, this was implemented in 2011 and standardized in 2018. On the other hand, telemedicine, which article 78 of the French law "*Hôpital, patients, santé, territoires*" (HSPT[1]) of July 21, 2009, recognized as a remote medical practice, mobilizes information and communication technologies (ICT). Although telemedicine does not replace current medical practices, it does provide a new method.

Another change is linked to the issues of individualization and patient empowerment in the health field. Marc Lemire shows that "the use of information and communication technologies is advocated to increase the involvement and responsibility of the population in health matters" [LEM 08, p. 3]. Other studies emphasize that the patient–doctor relationship has evolved over the course of history, the passive patient having become a fully-fledged actor in the decisions that concern them, or even completely involved in the care system, as shown by the current trend in therapeutic education, where patients – experts in their own illness – train other patients [JAF 15]. This movement is similar to that of the responsible or committed consumer, for whom information plays an important role when it comes to making decisions concerning diet (short food supply chains, organic products, etc.) [DUB 18].

1 Literally translated as Hospital, patients, health, territories.

The changes that have affected and continue to affect the field of health information are social, technical and documentary. They concern the production, publication and mediation of health knowledge, as well as information practices.

In terms of the general public's practices, information sources have diversified and, in addition to health professionals and their families, online resources, particularly health-related social media, have taken on an essential role the ways in which citizens obtain information, whether they are ill, close to the ill person or simply seeking information related to health issues. These devices, commonly considered as participative and social [COU 12], give access to hybrid information that mixes experiences with contents based on scientific data or practical information [CLA 10], [PAG 11].

In terms of the organization of knowledge in health systems, the "opening up of health information" on the Internet [MEA 20] has led both to a dispersion of information sources – linked to the loss of control over the supply of information by medical and scientific bodies – and to a heterogeneity of knowledge both in terms of modes of representation and content. Thus, images, whether fixed or animated in the form of videos, are developing for both health professionals and the general public.

After presenting the theoretical framework of our work, this chapter reviews more than 10 years of studies on health information, discusses the main results and shows how new information-seeking and knowledge-organizing practices are contributing to the evolution of health information.

2.2. Theoretical framework

The research presented in this chapter is part of the field of specialized information. All of these works have in common the adoption of a "contextualized" approach, that is, an approach which is based on a detailed analysis of the context in which the information is used (professional purpose, users' practices) [PAG 16] and of the content of the documents approached in their textual and discursive dimension [CLA 14]. Thus, our research is in line with the continuity of documentation approaches that define knowledge according to its situated uses.

Studies on the ways in which health professionals and patients inform themselves are part of the field of information seeking, and we consider information practices in the sense of Stéphane Chaudiron and Madjid Ihadjadene [CHA 10]. Information is understood as a process that leads an actor to seek to inform themselves, often with a defined objective and in a precise context [PAG 16]. We distinguish, in a rather classical way, professional context from the context of everyday life. We consider that informational activity depends on the circumstances in which it occurs, following Lucy Suchman's theory of situated action [SUC 87].

In the field of health, the medical professions evolve in a professional context, characterized particularly by the need for information related to their care and possibly teaching and research activities. Moreover, health professionals are required by law to update their knowledge. For patients, and more generally for citizens, the context is that of "daily life". They are looking for information related to their illness, that of a loved one or to keep abreast of current events, particularly when a health issue is covered by the media. Examples include the controversies surrounding vaccination or the Covid-19 epidemic.

In the approach we are developing, knowledge organization is linked to information practices and we consider these two fields to be complementary. At the beginning of our research into health information, we were interested in the knowledge extracted automatically to feed information systems. This is the case for a study we conducted in 2006 with the Grenoble University Hospital, aimed at automatically extracting knowledge based on sectorial classification systems, such as MeSH, in order to feed a bibliographic database intended for the hospital's doctors[2]. The objective was to streamline the information retrieval practices of healthcare professionals through the automated provision of journal articles, with physicians being asked to annotate these documents and share their comments. This techno-centric approach demonstrated its limits insofar as the practices of physicians had not been taken into account and they were not ready to participate in the production of medical knowledge that would form part of the information system.

2 In 2006, the University Hospital of Grenoble (CHUG) asked GRESEC to conduct a study on the information practices of the hospital's physicians. This study was part of a European research program (NOESIS) aimed at developing a module to help search for medical information on a "knowledge management" platform.

Gradually, we have adopted a broader definition of the organization of knowledge, including the social and technical conditions of knowledge dissemination, and have collected knowledge in the contexts of information use and mediation. Thus, we are currently interested in the forms of re-use of information in the diversity of professional situations (in or out of consultation), in the variety of health professions (medical or paramedical sector), in all specialties (specialists or general practitioners) or places of practice (private practice or hospital). We consider knowledge not only in its semiotic dimension but also by considering the transmission media (files, notebooks, post-its), the forms of codification (written notes, keywords, diagrams, etc.) and the devices (databases, personal sites, cloud storage spaces, etc.). Thus, the production and organization of knowledge are shaped by practices.

The study of information practices in health is a field of research in its own right, referred to as information seeking in health, which is highly developed in the United States and, to a lesser extent, in the Scandinavian countries. The studies most often focus on specific groups: young people [HIR 14, STJ 18], the poor [BAR 11], mothers [LEE 18], people suffering from certain diseases, etc., and their ways of obtaining information. The question of the credibility of information, relevance, trust and validation of sources is the subject of research [JOH 15, ORO 18]. The practices of professionals, most often doctors, are usually studied to model their information behavior or to ascertain how their practices have changed with the introduction of online resources [BEN 04, LEE 16]. The organization of knowledge in this field of research is not central but is always posed in terms of knowledge deficit (to access, understand or mobilize resources) or a lack of key words in certain languages.

2.3. Discourse collection and corpus analysis: approaches to health information from the "field"

Since 2005, we have approached what we consider to be "the terrain of health" in three ways. The first approach is that of the information practices of health professionals; the second is that of the info-communicational devices that are the social media of health and the third, which is a little more decentralized, is centered on the place that health occupies in society and its influence on information practices, as well as on the organization of knowledge.

Two types of methodologies were used: corpus analyses and speech collection (interviews and questionnaires). The corpus analyses were mainly focused on discussion forums (the *Doctissimo* forum and the *Sida*[3] *Info Service* forums) and personal health blogs since 2005. Although there were already studies on forums, especially in language sciences, most of them focused on a particular aspect, such as conversational analysis [MAR 04] or gave rise to a thematic analysis. Our approach, which is more comprehensive, considers the systems as a whole, taking into account their complexity and their editorial specificity in order to qualify the information available on the forums, to analyze the exchanges and micro-exchanges and, if necessary, to observe their place within the care or prevention system. Furthermore, social media is studied in its informational dimension and is understood as a source of information for patients and their relatives. Finally, we assume that we can understand practices through the traces of discourse, especially on sensitive subjects such as health, where it can be complicated to conduct interviews and to get in touch with sick people [CLA 17].

Second, we conducted field studies based on interviews or questionnaires with health professionals or forum moderators. For the health professionals, the aim was to collect information about their practices in the context of their professional activity. For the moderators, the objective was to validate the results obtained from a corpus analysis in the context of a study on HIV-related discussion forums[4].

2.3.1. *Study of the information practices of health professionals*

Several surveys on the information practices of health professionals were conducted between 2006 and 2018 among hospital physicians [STA 06, PAG 17], private physicians [MOU 09, PAG 17] and dieticians [CLA 19] with recurring questions: How do they get information? What are the preferred sources of information? Do they share information between peers? Does patient information via the Internet change the patient–practitioner relationship?

The views of health professionals were collected through semi-structured interviews during the study, conducted in 2005 at the University Hospital

3 "Sida" being the French term for AIDS.

4 This second point is outside the scope of this chapter.

of Grenoble [STA 06], and in 2006 and 2007 with private practitioners [MOU 09] and by questionnaire from health professionals [PAG 17] and dieticians [CLA 19].

The studies conducted since 2005 on the information practices of professionals, whether they are based on a quantitative or qualitative methodology, allow trends to be identified.

First, the weight of the context on information practices. We consider the context to be based on a set of elements that interact with one another. These elements concern the actors, the multiple activities in which they are involved, the informational environment and, finally, the organizational environment in which these practices take place. Thus, the practices of healthcare professionals differ according to profession (doctor, physiotherapist, midwife, nurse, dietician, etc.), according to the practice modalities (private practice, hospital), the status (practitioner or practitioner–researcher) or the activity that guides the search for information (medical diagnosis, response to a patient, writing an article). However, the information practices of health professionals are, in all cases, shaped by the obligation to continue training and updating the knowledge that governs these professions. Thus, as stipulated in article 11 of the code of medical ethics: "every physician must maintain and perfect their knowledge; they must take all necessary steps to participate in continuing education activities".

The longitudinal perspective of our studies highlights that since 2005 there have been recurring characteristics that mark a form of continuity over the last 15 years.

First, the information practices of health professionals cover various activities: research, production, sharing, valorization and dissemination of information, which are integrated into their daily work activities. They thus appear to be essential to their work but are not very quantifiable and ultimately not very visible [HUV 06]. In the case of doctors, they essentially surface in the interactions of professionals with their colleagues and patients [ISA 16].

Doctors and health professionals use information for a variety of purposes, including medical practice (diagnosis and treatment, responses to patients), scientific research (to update their knowledge, prepare for a conference or write an article) and administrative activities (managing patient files, reimbursements, relationships with health institutions and monitoring professional networks). To meet these requirements, professionals must consult scientific documents or data, information related to therapeutic treatments, prevention and public health policies and finally, for some, they need to know about the information shared on the Internet by patients or patients' relatives with a view to improving knowledge.

The information practices of health professionals are based on a fairly classic pattern that favors validated sources of information, disseminated by traditional and well-established channels (publishers, academic societies, pharmaceutical laboratories, etc.) and, contrary to other disciplines, leaves little room for open-access sources or information disseminated via professional or academic social networks.

The vast majority of the professionals we interviewed go directly to a source document when they are looking for information. Medline, the Vidal dictionary and the journal *Prescrire* were the first sources cited. Each profession then has its own specific sources: physiotherapists, nurses and midwives have specialized journals and books that they consult regularly, often on a daily basis. Professional or scientific journals related to the specialty (ophthalmology, hand surgery, obstetrics, etc.) are consulted, with academic societies playing an important role in disseminating information, updating research data and updating knowledge.

Within this traditional approach, the medical Internet has, of course, taken on a significant role. Thus, information-seeking practices have been renewed and professionals use the Internet either to access validated resources that were previously accessible elsewhere (databases, specialized journals), or to consult new resources available on the Web; resources emanating from public authorities, health organizations, associations or private actors. In particular, the sites of public institutions (such as the *Haute Autorité de Santé*, French High Health Authority), academic societies or scientific portals are mentioned.

Health professionals access the documentation that is useful to them in different ways. First, the vast majority use search engines. Google is the only general-purpose search engine cited, across all professions. Google Scholar, a scientific search engine, is cited only by the doctors at the university hospital.

Scientific databases are mentioned second, mostly by hospital doctors. The Pubmed database is the most frequently mentioned, followed by the ScienceDirect portal.

As early as 1997, Haug's study [HAU 97] on physicians' sources of information concluded that, in addition to obtaining information from the medical literature, practitioners frequently consulted their colleagues for answers to clinical and research questions. Indeed, regardless of their profession, the practitioners we met viewed their colleagues as sources of information. This finding confirms the importance of interpersonal exchanges in the professional setting. We have already shown that making use of colleagues when looking for information in a work context is very common because it often saves time and provides assurance in the reliability of the information transmitted [PAG 12]. A study published in 2010 [SEJ 10] showed the importance of the professional network of general practitioners, a network created on the basis of a, sometimes anonymous, relationship of trust which creates an ideal context for debate, confrontation of opinions and exchange of information.

Finally, medical sales representatives are the last-mentioned. In our studies, only doctors mentioned representatives of the pharmaceutical industry; these were both general practitioners practicing in private practice and specialists practicing in hospitals. Of the 25 doctors interviewed by Elodie Léon [LÉO 14], 10 refused to receive visits from medical sales representatives. The youngest doctors said they had never met with them, whereas the oldest doctors said they had stopped meeting with them about 10 years ago. It is the reliability of the information provided that is most often questioned. However, those who obtain information from the laboratories' documentation after visits from their representatives believe their critical mind can distinguish between commercial and scientific discourse.

In the introduction to this chapter, we distinguished four categories of health information. These different categories bring together sources of information mobilized by professionals, to varying degrees, in the context of their informational activities. Specialized medical information is consulted in order to update scientific knowledge; professional information regulated by public authorities is part of patient care; public health information is used to monitor current prevention and public health policies; finally, information "*on everyday health*" intended for non-specialists, which has developed considerably with the Internet, is used by certain professionals who seek to know and understand the concerns of their patients.

The information sources likely to be used by health professionals are numerous and varied. Since they have little time to obtain information in view of their daily activities, and the increasing importance of administrative tasks, they tend to favor one or two reference sources which they consult systematically, choosing to broaden their scope when their initial approach has not been successful.

2.3.2. *Studies related to the organization of health knowledge*

The organization of knowledge as an "object for the study of cognitive processes and intellectual techniques that make it possible to classify, index, formalize and model reality" [POL 05] has always been developed closely with information practices.

We adopt two approaches for collecting knowledge. The first involves extracting knowledge from corpora made up of information sources consulted by individuals in the context of their social practices. The second approach involves knowledge that is "already there", mobilized by individuals in the context of their info-communication practices, and we try to identify it *a posteriori*, most often through interviews or observations. In both cases, the knowledge is recorded on media, which remains a condition of its study.

The first approach is detailed in this chapter; the second can be found in Chapter 7 of this book, which recounts an interview with a doctor of ophthalmology concerning his personal development of a medical knowledge updating device.

The first approach resulted in the largest number of publications between 2010 and 2018. They all focused on health social media [CLA 10, PAG 11, PAG 14, CLA 17, CLA 18, MER 18] and followed previous research on FAQs [MOU 04].

In the set of resources of the medical Internet, social media occupy a special place. Firstly, because they are well referenced by search engines, they provide information that is often not accessible elsewhere and they allow people to find support because they create (or give the illusion of creating) personal relationships between Internet users and, through questions and answers, can provide individualized information. These systems are very popular with Internet users[5]. Patients, relatives of patients, caregivers, people living with a disease and members of the general public interested in health issues frequently use these systems to obtain information and to share their stories and their experience of living with a disease.

We understand social media as a corpus intended to implement methods of content characterization. The underlying questions were the following: For what reasons do Internet users use social media when they seek information about health? What place do these devices occupy in the set of available sources? What types of information do they seek? What place is given to the "traditional" expert? Are there differences in the way information is structured according to the pathologies? What role does emotion play in discussion forums, particularly in the evaluation of information? A doctoral thesis was defended on this subject in 2013[6].

The first study focused on the analysis of 24 threads selected from the Doctissimo forum dealing with various subjects ranging from well-being to serious diseases. Then, after having observed that the longer or more serious the disease, the more specialized the lexicon becomes, we worked on a corpus of threads dealing with families of diseases, such as rare or orphan diseases, serious diseases such as cancer or AIDS, or threads in which the diseases have consequences on diet (type 2 diabetes, bariatric surgery, etc.). Apart from Doctissimo, we also studied forums moderated by

5 To speak only of Doctissimo, Médiamétrie reported 1.8 million unique visitors in 2000 [CLA 10]. There were 13.7 million visitors in 2019, but other health media have emerged since. Available online: https://www.mediametrie.fr/sites/default/files/2020-01/2020%2001%2009%20-%20CP%20Audience%20Internet%20Global_Octobre%202019_1.pdf.

6 BATTAIA C., L'émotion dans les forums de discussion : structuration et évaluation de l'information de santé, PhD thesis in information and communication sciences, under the supervision of Laurence Balicco and Viviane Clavier, Université Grenoble 3, France, 2013.

doctors (e.g. Atoute.org and ligue-cancer.net), forums and blogs specializing in Crohn's disease, social media entirely dedicated to chronic diseases (Carenity). Finally, we were able to access the entire archive of the *Sida Info Service* (SIS) forum as part of a research project on HIV-related risks, funded by the *Agence Nationale de Recherche sur le Sida* (French National AIDS Research Agency).

As far as methods are concerned, content analysis is the basis of our approach. The indexing unit is the message. Contrary to classical content analyses that focus on this theme, we made the decision to translate in informational terms interactional notions that lead to distinguishing between exchanges such as requests for information or contributions of information. All of our studies show that there are fewer requests (about 20%) than contributions (about 80%), a result that confirms the purpose of the forum as a source of information. Once the interactions have been typed, apart from the friendly exchanges, we distinguish three types of information: practical information and advice; scientific information, which refers to scientific sources or to the words of health professionals (general practitioners or specialists) and, finally, testimonials. It appears that testimonials are always the most important and amount to 50% of the exchanges, a percentage that has been confirmed in all of our studies. Next comes practical information and advice, then scientific information in more variable proportions.

Then, depending on the objective of the study, the methods vary. For example, we have studied the place of medical expertise within the forums [PAG 14]:

– within the context of moderation, which can be provided by physicians;

– in the sources cited, which may refer to medical documents (prescriptions, test results), scientific documents (articles, websites), popular programs (documentaries, TV shows) or books written by doctors;

– in the words reported in direct or indirect discourse, which highlight an axiology of distrust versus trust towards the medical profession and show how the care of patients was experienced, in a close or depersonalized manner.

We also looked at the notion of theme by analyzing threads that mention food [CLA 18]. This more classical approach was based on the development of a set of categories that characterize the dietary knowledge mentioned by Internet users in the context of a pathology (types of food, types of diets, etc.) as well as on the way in which the knowledge is organized (in the form of a list of approved or forbidden foods). We also showed that the relationship of the participants to food was strongly influenced by the norms communicated by the public authorities in charge of nutritional prevention.

Finally, in the study on the forums set up to support AIDS prevention actions, we conducted a discourse analysis on the perception of risk through the expression of emotion in the discussion threads. The notion of risk, according to the SIS association, "is a complex subjective construction at the crossroads of knowledge and emotions depending on the context" [MAR 17]. For this study, contrary to the previous ones, the relevant indexing unit to study emotion is not the message but the whole thread. This study, which takes into account the subjective dimension of perceived risk, is based on the work of linguist Raphaël Micheli [MIC 13].

This research has provided elements for the characterization of the discussion threads and is at the center of a collaboration with a group of computer scientists, specialists in learning, and researchers at the LIRMM laboratory of the University of Montpellier [MER 18]. Thus, all levels of content description as developed over the years, have made it possible to create an annotation grid intended to analyze the discourses related to questions of sexuality, consumption of illicit products, and to bring out risk behaviors as well as the factors related to risk taking. A sub-corpus of 4,481 messages was annotated by us[7] from the 226,252 messages provided by SIS, and fed into a platform designed to facilitate annotation. This sub-corpus was then used to train a classifier designed to classify messages semi-automatically (data-driven corpus) according to genre, level of expertise, informational level, type of risk and emotions expressed. Once the messages were labeled, a streamgraph visualization, combined with a hierarchical navigation tool, was developed by the learning researchers to visualize the annotated speeches over time [*ibid.*].

---

7 And those of Amal JAOUZI, doctoral student in information and communication sciences at GRESEC.

Thus, social media can be considered a resource for extracting knowledge for various applications (here an aid for prevention professionals moderating a forum). This approach is now widespread, including among health professionals who were very resistant to social media in the 2010s. This is evidenced by the numerous theses and studies in public health or medicine that use social media as a resource to collect data. This is the case with Adeline Abbé's thesis in public health and epidemiology, which relies on the analysis of Internet users' stories through text mining in Doctissimo forums to document a study on anxiolytics or antidepressants in mental health [ABB 16]. In the same vein, social networks are used as a complementary source for the study of adverse drug reactions [BAG 16]. Internet users' stories provide opinions or views on a particular vaccine, such as HPV [FED 18], or shed light on risk behaviors such as risks associated with suicide [LUX 12]. But more recently, due to the volume of data available on social networks, polling firms are supplementing their quantitative surveys with automatable qualitative methods inspired by Big Data[8].

We can, of course, question this practice, knowing that testimonials equate to only 50% of the exchanges, that the methods used are based on learning techniques that only exploit the lexicon, which does not allow us to grasp the theme (which goes beyond the scope of the message), and that the enunciator is not considered in the exchanges, nor are the effects of repetition within the micro exchanges.

2.4. Conclusion: towards a characterization of health information

These different studies lead us to reexamine the contours of health information and to look for elements that can be used to characterize it.

First, the notion of context appears to be an element in the characterization of health information. The studies we have conducted on

8 For Laurence Stoclet, Deputy CEO of Ipsos, "Social networks are a new and very interesting source of data, but their representativeness is not always guaranteed. Surveys will therefore remain an important source of data, but we are supplementing them with passive measures known as 'big data' (listening to the Web, for example, or applications that track surfing habits on mobiles, tablets or PCs), as well as qualitative techniques for observing or measuring emotions (facial coding, biometrics, etc.)". Available online: http://www.lenouveleconomiste.fr/lesdossiers/instituts-detudes-et-de-sondages-big-data-highlighting-you-tou-31828/.

information practices have always integrated the notion of context [PAG 16]. Thus, informational activities, observed in a professional setting, are influenced by a context that can be understood at three levels: micro (the individual and their daily life), meso (in relation to the organization) and macro (on the scale of society and the changes that affect it and also influence social practices) [*ibid.* paragraph 43]. In practice, this typology is difficult to apply when it comes to distinguishing between narrow (micro) and broad (macro) contexts, for example, technical changes are perceptible at both the societal and individual levels.

In our most recent research related to food and nutrition [CLA 18], we have worked more closely on the link between information practices and the identification of a context that can be described as broad. This context is the result of long-term changes in society, and these changes also have an influence on information practices and the mediation of knowledge. The societal context that interests us is marked by health issues. Thus, health, beyond the care of the sick and their medical care, has largely penetrated the public space and the media now play a central role [LAF 07a, LAF 07b, ROM 10], whether in the revelation of cases [MAR 10]; in highlighting the sick thanks to their testimonials on television programs [ROM 07]; in the involvement of informed and educated patients who participate in their own medical follow-up [MEA 20] or who can become partners in the therapeutic education of other patients [SAN 19], and in the communication of public prevention campaigns on health-related risks [ROM 13]. The strong media coverage of health issues in the public arena is not recent but accompanies the long process of medicalization of society that began in the 1950s, as described by the sociologists Pierre Aïach and Daniel Delanoë in 1998 [AIA 98][9].

In terms of content, health information is characterized by strong media coverage of medical knowledge through the large number of programs, reports and articles on health and well-being in the media: the place given to scientific experts in media discourse on the subject; prevention or education campaigns on health or nutrition; press releases from public authorities and health agencies or advertorials from the agri-food industry.

9 For more details, see the general introduction to the book.

Finally, health information is characterized by the profusion of actors who produce, edit and distribute content. These actors, when they do not come from the medical profession and do not publish in the scientific sphere, call on experts to validate and legitimize the content they produce. Thus, in the traditional media, the scientific guarantee is provided by public health institutions, health security agencies and doctors (sometimes very controversial). When it comes to food, dieticians, nutritionists, coaches or even chefs may be called upon. Some content from the food industry "mimics" scientific productions by using pseudo-experts to legitimize the health benefits of products. Finally, in health-related social media, the scientific guarantor is sometimes a media personality from the popular press or TV shows. Health professionals are also quoted by Internet users. These are most often their own doctors, general practitioners or specialists.

The place occupied by health in the public arena and the strong media coverage of medical knowledge have consequences on the information practices of individuals and on the organization of knowledge.

In terms of information practices, in this context where health appears to be a central concern, where individuals are invited to take charge of their own health and where normative directives concerning well-being or health are very strong, we have observed several trends:

– first, it appears that some patients are more sensitive to health issues than others. This is the case for women, vulnerable people or young parents when it comes to looking after their children's health;

– second, attitudes towards normative directives can be divisive. For example, with regard to vaccination or "eating a balanced diet", we observe strong rejection or adherence, attitudes that strongly influence information practices. Some people seek "alternative" information by rejecting public discourse or that of the traditional media; they share information via social networks and favor word of mouth. Others declare that they no longer seek information, citing a lack of confidence in the information found in any source.

With regard to information sources, the role that health plays in society also has consequences for the organization of knowledge and is reflected in writing formats, modes of semiotization and terminology. This propensity to mimic scientific discourse appears in advertising inserts, for example, that call upon nutritional knowledge; in discussion forums that describe the stories of patients suffering from this or that pathology and in the practice of "pre-formatted" quotes from experts that constitute "quick reaction kits" for commenting on current events[10]. Health professionals, for their part, are often led to "reframe" patient knowledge during their consultations, as knowledge often appears to be unsound or to contain many preconceived ideas, as is the case with diet. Generally speaking, the multiplication of medical specialties makes it necessary to follow networks of professionals and to update knowledge.

2.5. References

[ABB 16] ABBE A., Analyse de données textuelles d'un forum médical pour évaluer le ressenti exprimé par les internautes au sujet des antidépresseurs et des anxyolitiques, Public health and epidemiology, PhD thesis, Université Paris-Saclay, 2016.

[AÏA 98] AÏACH P., DELANOË D. (eds), *L'ère de la médicalisation*, Ecce homo sanitas, Éditions Economica, Paris, 1998.

[BAG 16] BAGHERI H., LACROIX I., GUITTON E. *et al.*, "Cyberpharmacovigilance : les réseaux sociaux sont-ils utiles en pharmacovigilance ?" *Thérapie*, vol. 71, no. 2, pp. 235–239, 2016.

[BAR 11] BARMAN-ADHIKARI A., RICE E., "Sexual health information seeking online among runaway and homeless youth", *Journal of the Society for Social Work and Research*, vol. 2, no. 2, pp. 89–103, 2011.

[BEN 04] BENNETT N., CASEBEER L., KRISTOFCO R. *et al.*, "Physicians' Internet information-seeking behaviors", *Journal of Continuing Education in the Health Professions*, vol. 1, no. 2, pp. 31–38, 2004.

10 At the initiative of the Ministry of Education and Research, the multi-year law on research, presented on September 21, 2020, to the Assemblée Nationale, establishes the creation of a House of Science and Media. This institution, which will function as a communication agency specialized in scientific and technical subjects, will offer journalists summaries and "rapid reaction kits", that is, ready-to-use kits of expert quotes commenting on current events and scientific controversies (cf. "L'information scientifique à l'ère du prêt à l'emploi", Le Monde, September 23, 2020).

[CHA 10] CHAUDIRON S., IHADJADENE M., "De la recherche de l'information aux pratiques informationnelles", *Études de communication*, no. 35, available at: http://edc.revues.org/index2257.html, 2010.

[CLA 10] CLAVIER V., MANES-GALLO M.-C., MOUNIER E. *et al.*, "Dynamiques interactionnelles et rapports à l'information dans les forums de discussion médicale", in MILLERAND F., PROULX S., RUEFF J. (eds), *Web social : mutation de la communication*, Presses Universitaires du Québec, 2010.

[CLA 14] CLAVIER V., L'organisation des connaissances au prisme du langage, du texte et du discours. Un parcours en recherche d'information, HDR thesis, Université Stendhal Grenoble 3, 2014

[CLA 17] CLAVIER V., PAGANELLI C., "Une approche méthodologique croisée du traitement des données de la recherche : le cas d'un corpus d'échanges issus de médias sociaux dans le domaine de la santé", *Colloque COSSI 2017, Méthodes et stratégies de gestion de l'information par les organisations : des "Big Data" aux "Thick Data", 85ème congrès de l'ACFAS*, McGill University, Montreal, 11 and 12 May 2017.

[CLA 18] CLAVIER V., "Le bon et le bien manger dans les forums de santé : des ressources pour s'informer sur l'alimentation", in HUGO-GENTIAL C. (ed.), *Bien et bon à manger. Penser notre alimentation du quotidien à l'institution de santé*, Editions Universitaires de Dijon, 2018.

[CLA 19] CLAVIER V., "La place de l'information dans les pratiques professionnelles des diététiciennes : au croisement des missions d'éducation, de prévention et de soin", *I2D – Information, Données & Documents*, no. 1, pp. 114–133, 2019.

[COU 12] COUTANT A., STENGER T., "Les médias sociaux : une histoire de participation", *Le Temps des médias*, vol. 18, no. 1, pp. 76–86, available at: 10.3917/tdm.018.0076, 2012.

[DUB 18] DUBUISSON-QUELLIER S., *La consommation engagée*, Presses de Sciences Po, Paris, 2018.

[FED 18] FEDERSPIEL A.-L., Quels sont les questionnements sur la vaccination contre le papillomavirus humains exprimés par les jeunes filles en âge d'être vaccinées sur les forums internet ? General Medicine thesis, Université Lille 2, available at: https://pepite-depot.univ-lille2.fr/nuxeo/site/esupversions/52bce079-7436-4d59-a889-305da5d1f8a2, 2018.

[HAU 97] HAUG J.D., "Physicians' preferences for information sources: A meta-analytic study", *Bulletin of the Medical Library Association*, vol. 85, no. 3, pp. 223–232, July 1997.

[HIR 14] HIRVONEN N., EK S., NIEMELÄ R. *et al.*, "Socio-demographic characteristics associated with the everyday health information literacy of young men", *Proceedings of ISIC, the Information Behaviour Conference*, Leeds, 2–5 September 2014.

[HUV 06] HUVILA I., The ecology of information work: A case study of bridging archaeological work and virtual reality based knowledge organisation, Dissertation, Åbo Akademis Förlag –Åbo Akademi University Press, Finland, 2006.

[ISA 16] ISAH, E.E., BYSTRÖM, K., "Enacting workplace information practices: The diverse roles of physicians in a health care team", *Proceedings of the Ninth International Conference on Conceptions of Library and Information Science*, Uppsala, Sweden, available at: http://InformationR.net/ir/22-1/colis/colis1650.html, June 2016.

[JAF 15] JAFFIOL C., "Éducation thérapeutique et patient expert. Repenser la médecine face au défi des maladies chroniques, invité-discutant", *Bulletin de l'académie nationale de médecine*, vol. 199, nos 4–5, pp. 597–599, 26 May 2015.

[JOH 15] JOHNSON F., ROWLEY J., SBAFFI L., "Modelling trust formation in health information contexts", *Journal of Information Science*, vol. 41, no. 4, pp. 415–429, 2015.

[LAF 07a] LAFON B., "Le cancer en *prime time*. Émergence d'une confrontation distanciée avec la maladie par les fictions télévisées", *Questions de communication*, no. 11, pp. 129–147, 2007.

[LAF 07b] LAFON B., PAILLIART I. (eds), "Malades et maladies dans l'espace public", *Questions de communication*, no. 11, pp. 7–15, 2007.

[LEE 16] LEE S.T., LIN J., "A self-determination perspective on online health information seeking: The internet vs. face-to-face office visits with physicians", *Journal of Health Communication*, vol. 21, no. 6, pp. 714–722, 2016.

[LEE 18] LEE H.S., "A comparative study on the health information needs, seeking and source preferences among mothers of young healthy children: American mothers compared to recent immigrant Korean mothers", *Information Research*, vol. 23, no. 4, available at: http://informationr.net/ir/23-4/paper803.html, 2018.

[LEM 08] LEMIRE M., "Application du concept de responsabilisation personnelle aux usages sociaux des technologies d'information et de communication en santé", *tic&société*, vol. 2, no. 1, available at: https://doi.org/10.4000/ticetsociete.351, 2008.

[LÉO 14] LÉON E., Les pratiques de recherche documentaire des médecins généralistes : les freins et les difficultés pour l'accès à une information de qualité, Human Medicine and Pathology thesis, Université de Bordeaux, 2014.

[LUX 12] LUXTON D.D., JUNE J.D., FAIRALL J.M., "Social media and suicide: A public health perspective", *American Journal of Public Health*, vol. 102, no. 2, pp. 195–200, 2012.

[MAR 04] MARCOCCIA, M., "L'analyse conversationnelle des forums de discussion : questionnements méthodologiques", *Les Carnets du CEDISCOR*, no. 8, pp. 23–38, 2004.

[MAR 10] MARCHETTI D., *Quand la santé devient médiatique. Les logiques de production de l'information dans la presse*, Presses universitaires de Grenoble, 2010.

[MAR 17] MARCE F., "Les appels à Sida Info Service, quelles perceptions pour quels enseignements ?" Journées PrEP/CeGIDD, Une formation de la Société Française de Lutte contre le SIDA, available at: http://www.sfls.aei.fr/ckfinder/userfiles/files/Formations/pdf/2017/Cegidd_Franck-Marce.pdf, 2017.

[MEA 20] MEADEL C., "Internet user-patient(s), a collective adventure", in PAILLIART I. (ed.), *New Territories in Health*, ISTE Ltd, London and John Wiley & Sons, New York, 2020.

[MER 18] MERCADIER, Y., AZE J., BRINGAY S. *et al.*, "#AIDS Analyse Information Dangers Sexualité: Caractériser les discours à propos du VIH dans les forums de santé", *29es Journées francophones d'ingénierie des connaissances*, IC 2018, Nancy, France, pp. 71–86, 2018.

[MIC 13] MICHELI R., "Esquisse d'une typologie des différents modes de sémiotisation verbale de l'émotion", *Semen*, no. 35, available at: http://semen.revues.org/9795, 2013.

[MOU 04] MOUNIER E., PAGANELLI C., "The representation of knowledge contained in technical documents: The example of FAQs", *Knowledge Organization and the Global Information Society, Proceedings of the 8th International ISKO Conference*, London, July 2004.

[MOU 09] MOUNIER E., PAGANELLI C., "Pratiques et usages de l'information spécialisée chez les médecins spécialistes : place et rôle des structures documentaires", *Congrès international ACSI (Association Canadienne des Sciences de l'Information). Tracer l'horizon informationnel du XXIe siècle : frontières, passerelles et carrefours*, Ottawa, available at: http://www.cais-acsi.ca/proceedings/2009/Mounier_Paganelli_2009.pdf, 28–30 May 2009.

[ORO 18] OROSZLÁNYOVÁ M., TEIXEIRA LOPES C., NUNES S. *et al.*, "Can user and task characteristics be used as predictors of success in health information retrieval sessions?", *Information Research*, vol. 23, no. 3, paper 801, available at: http://InformationR.net/ir/23-3/paper801.html, 2018.

[PAG 11] PAGANELLI C., CLAVIER V., "Le forum de discussion : une ressource informationnelle hybride entre information grand public et information spécialisée", in YASRI-LABRIQUE E. (ed.), *Les forums de discussion : agoras du XXIe siècle ? Théories, enjeux et pratiques discursives*, L'Harmattan, Paris, 2011.

[PAG 12] PAGANELLI C., Une approche info-communicationnelle des activités informationnelles en contexte de travail : acteurs, pratiques et logiques sociales, Information and Communication Sciences PhD thesis, Université de Grenoble, 2012.

[PAG 14] PAGANELLI C., CLAVIER V., "S'informer via des médias sociaux de santé : quelle place pour les experts ?", *Le Temps des Médias*, no. 23, pp. 141–143, available at: http://www.histoiredesmedias.com/S-informer-via-des-medias-sociaux.html, winter 2014.

[PAG 16] PAGANELLI C., "Réflexions sur la pertinence de la notion de *contexte* dans les études relatives aux activités informationnelles", *Études de communication*, vol. 46, no. 1, pp. 165–188, available at: https://www.cairn.info/revue-etudes-de-communication-2016-1-page-165.htm, 2016.

[PAG 17] PAGANELLI C., "Informations de santé et pratiques informationnelles des médecins : quelles approches en SIC?", in FRAYSSE P., GARDIES C., FABRE I. (eds), *Sur les sciences de l'information et de la communication : contributions hybrides autour des travaux de Viviane Couzinet*, Editions Cepadeus, Toulouse, 2017.

[POL 05] POLITY Y., HENNERON G., PALERMITI R. (ed.), "L'organisation des connaissances. Approches conceptuelles", *4th ISKO-France Conference Proceedings*, 3 – 4 July 2003, L'Harmattan, Grenoble, 2005.

[ROM 07] ROMEYER I., "La santé à la télévision : émergence d'une question sociale", *Questions de communication*, no. 11, pp. 51–70, 2007.

[ROM 10] ROMEYER I., *La santé dans l'espace public*, Presses de l'EHESP, Rennes, 2010.

[ROM 13] ROMEYER H., MOKTEFI A., "Pour une approche interdisciplinaire de la prévention", *Communication & langages*, vol. 176, no. 2, pp. 33–47, 2013.

[SAN 19] SANNIE T., "Des patients partenaires d'enseignement : une avancée historique dans les formations médicales", *Santé Publique*, vol. 31, no. 4, pp. 473–474, 2019.

[SEJ 10] SEJOURNE E., Pare, F., Moulevrier, P. *et al.*, "Modalités de constitution du carnet d'adresses des médecins généralistes", *Pratiques et organisation des soins*, vol. 41, no. 4, pp. 331–339, 2010.

[STA 06] STAII A., BALICCO L., BERTIER M. *et al.*, "Les pratiques informationnelles des médecins dans les centres hospitaliers universitaires : au croisement de la logique scientifique et de la culture professionnelle", *Revue canadienne des sciences de l'information et de bibliothéconomie*, vol. 30, nos 1/2, pp. 69–90, March–June 2006.

[STJ 18] ST. JEAN B., GREENE TAYLOR N., KODAMA C. *et al.*, "Assessing the health information source perceptions of tweens using card-sorting exercises", *Journal of Information Science*, vol. 44, no. 2, pp. 148–164, 2018.

[SUC 87] SUCHMAN L., *Plans and Situated Actions: The Problem of Human/Machine Communication*, Cambridge University Press, 1987.

3

Health Literacy in Complex Digital Information Environments

3.1. Introduction

Health literacy must now be considered and studied, taking into account digital environments as well as the information on offer that users and patients may encounter online. Work in the field of health literacy has been ongoing for some 40 years. The concept appeared in 1974 [SIM 74], the same year as the concept of information literacy [ZUR 74]. Similarities between the two can be seen both in terms of definition and content, particularly because the main issue in health literacy is information. The concepts of preventive medicine or therapeutic education are not totally removed from the aspects covered by the term. Health literacy appears to be a broad and all-encompassing concept that is not limited to medical literacy. Indeed, it concerns users who are not necessarily patients [PEE 09, SAU 09]. Informational and digital health practices are now common and are also performed outside of care pathways. However, patient community sites and forums are becoming better recognized aspects in healthcare pathways.

This chapter thus seeks to show the evolution of the concept and its realities in relation to the transformation of the supply of health information. We want to understand the relationships between health literacy and other literacies that appear to be closely related: information literacy, digital literacy, information literacy and digital culture [BAW 01, LED 11], etc.

Chapter written by Eloria VIGOUROUX-ZUGASTI, Olivier LE DEUFF and Amar LAKEL.

Indeed, increasingly it appears as a cross-disciplinary literacy or transliteracy [THO 07], combining several types of literacy (informational, media, health, etc.). Therefore, the question of the level or measurement of this literacy seems complex to assess, despite assessment instruments that try to take into account as many criteria as possible. Like many other literacies, health literacy appears to be regularly in crisis [GRA 87]. It sometimes gives rise to concern on the part of national or international institutions, especially since it is not only an intellectual or cultural literacy but a major public health issue. Recent definitions try to take these parameters into account. The 1998 definition shows its importance: "Health literacy represents the cognitive and social skills which determine the motivation and ability of individuals to gain access to, understand, and use information in ways that promote and maintain good health"[1]. It is in this broad approach to health literacy – which we will now refer to as "digital health literacy" – that our scientific approach is based.

In the first part of the chapter, we open a discussion on the issue of global mixed methods that we have mobilized in research projects on health information on the Internet, particularly in the context of our participation in the EUR "Bordeaux Public Health"[2]. In a second step, we review the results of our surveys on the web dimension of health research and on the connected uses of seniors dedicated to this topic. The results will be based on quantitative data from web analysis, as well as on qualitative data from interviews with seniors. Finally, we show the benefits of conceiving digital health literacy as a literacy dedicated to considering health information on the Internet as an important factor of therapeutic education, which is currently at the heart of public health policies.

3.2. Analyzing health information on the Internet: the need for a mixed and comprehensive approach

Understanding the process of influence on the Internet is a major but complex issue for social science research. The object itself, despite a century of research, is still poorly defined. Influence is the effectiveness of an act of

1 WHO Health Promotion Glossary. World Health Organization, Geneva. https://www.who.int/healthpromotion/about/HPR%20Glossary%201998.pdf?ua=1.

2 My Web Intelligence in health information project (2014–2017, Conseil Régional – Direction du Numérique), Information Risk in Seniors and Self-Medication project (2015–2016, Bordeaux Montaigne University and MAIF Foundation).

power in that it aims to act on the actions (or representations) of another person [LAK 08]. Research on the side of the influencers (i.e. on their capacity to influence) must take into account the social characteristics of the speaker, their communicative practices, the social system in which they act and their role in hierarchical networks of actors. On the side of the influenced, it is research on literacies as a key concept that allows us to understand the learning and subjectivation paths. Influence and Literacy are two sides of the same coin that aim to study the Web as a medium of transformation of informational practices.

3.2.1. *Approaching public health information literacy through the lens of information quality in ICS*

The first approach to the question of health information on the Internet, from the point of view of the ICS, involves studying the structure of the offer, aiming to set up the conditions for a sociology of digital transmitters and even influencers. An examination of the scientific corpus that studies the supply of health information, particularly on the Web, shows that a high proportion of articles come from Public Health research (Social Sciences and Health) and that the Humanities and Social Sciences, in general, only deal with this subject in less than a third of the corpus (ICS journals are almost not represented) [LAK 19]. This corpus can be classified into three main scientific approaches. Firstly, for public health researchers, it is a matter of examining the quality of content with respect to the norms of medical discourse by giving (bad) marks to online content. A second approach, based on a sociological approach, analyzes the rise of the patient's voice, which tends to redefine the traditional discursive economy in Web 2.0 spaces (blogs, forums, social networks, etc.).

As for the Public Health approach in ICS, it specifies this research axis by insisting on the new economies of content, by developing an analysis of discursive regimes in the framework of a critique of the cultural industry. Much more than a sociology of actors, the challenge here is to pay attention to the sociosemiotic markers of discourse.

The first line of research, in public health (section 46 of the *Conseil National des Universités* (French National Council of Universities)), attempts to evaluate digital corpora with respect to the standards of medical publications on a scale of scientific quality. The quantitative approach is very formalized and quite stable from one article to the next. This methodological consistency is strongly conditioned by the disciplinary homogeneity which leads to the same analysis models being repeatedly applied to different corpora. The idea is therefore to find the doctor's voice on the Internet and to evaluate its quality with regard to the standards used in health (academic societies, medical recommendations, etc.). We therefore naturally exclude patient forums and blogs whose discursive regime is clearly considered by the public health researcher as irrelevant, even though it is prolific on the Web.

Researchers then rely on objective and institutional evaluation grids such as DISCERN, HON, NetScoring, or even the standards of the European Commission. In most cases, researchers combine these standards to propose their own evaluation grid [ESP 14, FRE 01]. The results of the analysis are not surprising. Non-medical corpora are not "good" medical corpora. The presence of the medical profession very often exceeds half of the corpus, and patient associations and patients themselves are barely represented (around 10% in most articles). Half of the discourse is addressed to professionals, and in more than half of the cases, the author is not identifiable, nor the date, nor the updates. The authors quite often recognize that good bibliographic references are present but often insufficient. For two-thirds of the pages analyzed, they are considered mediocre if not bad. The actors recognize that there is very little dangerous information but that the corpus is overall quite poor. Added to this, longitudinal studies show a high level of instability of the corpus from one year to the next with more than half of the corpus disappearing from search engines for the same query. When it is not clearly judged to be of no interest at all [ESP 13], the words of the patient and their expectations are described as having little interest in terms of medical legitimacy.

The second line of research, stemming from digital sociology and ICS, is based on a qualitative approach to online discourse that is close to the paradigm of pragmatic interactionism. If the corpus proves to be neither exhaustive nor representative, these "exploratory studies" remain cautious as to the representativeness or even the generalization of their observations because of the impossibility of constituting either a reference population or a representative sample [LAK 19]. By relying both on the formal description of interaction sequences based on the Goffmanian model and on a lexical analysis of thematized discourse, the approach of this research focuses on the emergence of new discursive strategies of patients and caregivers [CON 03, STAI 10, PAG 14]. All of this research seems to stabilize in a field of hypotheses distributed on three levels: strategic, pragmatic and stylistic [BAT 13, PAG 14]. On the strategic level, it is the patient's voice that seems to be the focus of all attention. The research in SHS then focuses on the emergence of a new discursive regime that is very present in these digital spaces and that gives them their own specificity. It is a regime of personal testimony, of individual lived experience, which underpins the specificity of ICS approaches.

Faced with these two approaches, we wanted to develop an open quantitative approach based on the development of a tool for extracting (a crawler), archiving and qualifying the Web for digital studies, which can remove the constraints currently weighing on the formation of online corpora: My Web Intelligence[3]. The objective is to provide all experts and researchers who wish to develop studies in the field of digital intelligence, digital humanities and sociology of the Web, with a simple device for building a web corpus representative of an open controversy. Establishing digital corpora for social science research is then at the heart of what we have called the tests of digital methods [COR 19].

From queries finding a semantic dictionary on a disease, we obtain, from two search engines (Google.fr and MSN.fr), 1,235 relevant pages that we group in a second data set at the level of expression domains[4]. In this research, we start from the usage scenario of an average French Internet user asking about a disease. The terms were chosen to be neutral and short, in order to cover the most open sample possible[5]. In this research, we wanted to avoid the sampling bias introduced in previous studies by a socially situated vocabulary. As we have seen, the choice of a socially situated vocabulary necessarily over-represents the populations usually using these keywords.

3 *My Web Intelligence* is today a prototype platform created within the MICA that runs in a Python environment on an SQLite database: http://mywebintelligence.net.

4 The expression domain is not quite the domain name because in the case of content farms (social networks, blogging platforms, etc.), we had to develop heuristics to split these domain names into sub-domains corresponding to editorial units (pages, profiles, groups, etc.). In this study, we retained 846 domains of expressions.

5 Our research aimed to study web pages addressing the topic of asthma in children using the words "*asthme*, asthma" associated with "*jeune*, young", "*enfant*, child", "*bébé*, baby", "*nourrisson*, infant", "*infantile*, infantile" and "*adolescent*, teenager". For more details on the methodology, see [LAK 19].

Technical words, scientific or familiar words, affective words, the lexical field chosen in the dictionary of queries is necessarily a sampling bias. We measure that the meaningful Web for a user of these search engines is a limited Web because it is formally structured by the ranking algorithms of these infomediaries, socially structured by the marketing strategies of the speakers and intentionally structured by the receiver's query. *In fine*, these three rules lead to the creation of a corpus biased by the semantic field used in the retrieval requests.

3.2.2. *My Web Intelligence: a model for analyzing health information controversies*

The first level of analysis is therefore the page containing a statement about asthma in children. A first automatic extraction obtains the complete HTML code, the title, the URL and the description contained in the meta tags. Then, algorithms ensure a first layer of extraction by detecting the main content of the document (readable); the URLs of links (urls), images (imgs) and videos (videos) at the heart of this content. The goal is to build the document's citation network. An extraction by crawling at several depth levels enriches the corpus. The depth variable reminds us if the document was obtained at the start (level 0) or by extraction (level n).

In addition to this first level of extraction, there is an automatic level of external extraction via Web services by API. At this stage, we find the measurement of the social resonance measurement of the URL (*facebook_like*, *facebook_share*, *linkedin_share*) which remains to this day the only global indicator of a page's impact. Indeed, the first limitation lies in access to information that measures the use of a piece of content. Between the beginning of the project and its end (18 months), we saw all the social networks close their API doors and more and more indicators were no longer available. We were only able to keep the *likes* and *shares* on Facebook. This problem highlights one of the major limitations of *digital methods* that access data owned by private companies that play an arbitrary role in controlling access to their resources for research purposes [COR 19]. In a third step, human annotation allows us to contextualize the document thanks to three variables: the publication date (*publication_date*) on the Google search engine, the media type (*media_type*) and the emitter type (*emmitter_type*), which we have deliberately kept open.

Finally comes the set of structural network analysis algorithms (Social Network Analysis) from the hyperlinks of internal citations in the corpus [SCO 17, WAS 94]. For each URL, the orientation (indegree and outdegree) and weight (weighted degree, weighted indegree, weighted outdegree) of the links (degree) have been calculated according to the implementation of the Gephi software. The ModularityClass community detection/qualification algorithm [BLO 08] was used to detect the communities of citations and classify the pages and nodes into citation communities. We also computed at the level of each domain the betweenness centrality and the local Page Rank [BRI 05]. The Page Rank is calculated only at the level of the sites present in the corpus. This allows us to know the strength of the presence of a node at the heart of a network by calculating the probability of encountering this node when randomly moving in the network. Finally, Authority Rank, which allows us to evaluate and rank domains according to their authority in the network, and its corollary Hub Rank, which allows us to detect the domains of expression that play the role of infomediaries (or watchdogs) at the heart of the network by multiplying the links to a given community [KLE 99], allow us to calculate the authority of the nodes within constituted communities. Finally, closeness centrality calculations, and in particular harmonic closeness centrality [BRAN 01, BRAN 05], allow us to calculate the nodes that are at the heart of the network and best placed to meet the other nodes (calculation of the average distance to each node.

We wanted to make the Network Actor Theory [LAT 07, LAT 10, LAT 14] operational at the level of digital media studies [BAR 17, BOU 16]. Finally, it is necessary to underline the rarity of the phenomenon of social resonance throughout the corpus. The measures of social resonance (likes, shares, comments, etc.) allow us to understand the reception of a unit of information. If we add up the structural analysis indicators of the citation network, we can build a multivariate influence scale that quantifies the notion of influence [JEN 16, ROS 17]. Added to the factor of position (who speaks, about what and on which device), we obtain three factors allowing us to study the strategies of actors at the heart of a digital sociology of controversies. The mapping of actors can then lead to studies of the "places" of influence of knowledge on the Internet by adding to a description of the positions taken, their social impact and their socialization environment.

The examination of Web corpora containing health information highlights an informational complexity and influence games specific to the structure of the Web that contrast with the apparent ease of use of the search engine. In the study of our corpus, the likes and shares of social networks all have a statistical distribution law in the form of a strong decreasing exponential (or Power law [BAR 14, BAR 15]). A hierarchy emerges from the library of documents; a determining maze of navigation is revealed from the citation paths, and informal editorial communities show “small worlds” that are more or less endogenous. The Web is not a neutral list of links displayed on a search engine; it is a complex social informational architecture that determines the subjectivation of the reader. It then becomes necessary to couple the study of the corpus with that of the uses of this information, by tracing the effective wanderings of the Internet users and the process of integration of these paths.

3.2.3. *Qualitative approach: understanding the daily use of digital health care*

In parallel to the structural analysis of the Web, we propose a health information approach that completes the previous results with a longitudinal qualitative survey on online health practices, among a retired population, aged 60–80, using digital technologies. The choice of the qualitative method aims at understanding the health practices of elderly Internet users, both in terms of searching for connected information and in terms of self-medication, applied to the users themselves and to the people around them, for whom they are doing the research. Although the choice of the population studied may seem surprising, it is ultimately relevant in the context of an approach to the digital health literacy of users. Indeed, this population, under the effect of biological aging, is particularly interested in the field of health and its management by the principle of self-evaluation [FOU 01]. It also presents complex forms of acculturation (digital, media and digital) [VIG 17], little studied in human sciences, in particular in French CIS. Risks in public health appear to be obvious, especially since health research online can lead to self-medication, more or less well controlled according to the capacities of access to relevant information [*ibid.*].

Our qualitative approach sought to better understand the modalities of reception of digital information and communication technologies among the population studied, since “confrontation with the field is the condition for the perception of social practices” [MON 08]. In addition, it also seeks to study the implementation of the concept of digital health literacy in a population of users. The idea is then to verify the correlation between this appropriation and the development of good or bad online health practices, applied to ourselves or to a third party.

Our qualitative approach includes two parts. The first is part of the project *Risque Informationnel Chez les Senior et Automédication* (RICSA), carried out by Bordeaux Montaigne University and the MAIF Foundation for Research and Risk Prevention in 2015 and 2016. The selected sample includes 31 seniors, interviewed in individual semi-structured interviews. The respondents are over 60 years old, selected according to multiple criteria: age, gender, place of living, standard of living, familiarity with digital technologies, etc. The sample was compiled according to a quota method, with a desire for equal distribution between the different criteria retained (age, place and standard of living, gender, former profession, social environment).

We deliberately chose to structure the sample according to a principle of heterogeneity. We sought to diversify the contexts of online health use, in order to understand the convergent and divergent nature of the behaviors observed, as well as the complexity of the relationship to digital literacy. In addition, we felt it was important to understand "the complexity of the practices analyzed [through the investigation] of forms of knowledge other than that held by the researcher" [MON 08]. This approach seemed relevant to our approach. It allows us to encourage the emergence of concrete data that can contribute to deepening and structuring the concept of digital health literacy, without being too attached to a particular scientific paradigm. We then conducted a new wave of interviews in 2019, three years after the first phase, carried out with 30 of the participants of the RICSA project, using the same survey and analysis grid, with the same research objectives. Through this longitudinal study, we wish to verify the evolution of online health uses and, in particular, the evolution of the relationship to digital health literacy.

Concerning the analysis of the qualitative survey, we first chose the methodology of continuous thematic attribution, i.e. readings of the data seeking to invest, through short expressions, the substance of the analyzed statement [PAI 08], in order to link it with the other data. This technique has the advantage of allowing a certain increase in the generality of the field data. The thematic analysis "has as its essential function neither to interpret [...] nor to theorize [...] nor to bring out the essence of an experience", but "serves to identify and summarize the themes present in a corpus" [PAI 12].

We then proceeded to a categorical analysis of the content, in addition to the thematic analysis, based on the fact that "the aim [of content analysis] is to identify several categories [which] will allow the researcher to classify and order the whole corpus" [BAR 12]. This methodological contribution thus allows us to go beyond the relatively pragmatic thematic sorting, in order to bring the analysis closer to scientific exploitation of the data.

3.3. Key findings and issues in digital health literacy

The results obtained allow for a better understanding of the state of the information on offer, as well as the editorial issues that are at stake. The need to increase the skills of users appears all the more necessary.

3.3.1. *Strategic positions of health mediators*

In the study of the typology of speakers, we note that the discourse is dominated by health actors and journalists (two thirds of the sites). Patients and companies form the final third of the actors in this field. "Alternative" discourses[6] play a very minor role (about 1 or 2% of the sites). The most used communication platforms are editorial sites with a strong strategy of periodic contents (1st third), institutional presentation sites (2nd third) and then Web 2.0 platforms such as social networks, forums and blog platforms (last third). This introductory description reveals very typical and fairly intuitive digital spaces. However, it is necessary to qualify this description for healthcare actors because the associative form of healthcare practitioners generally aims to develop a strategy of information resource portals as much as an institutional strategy. And practitioners communicating individually have a social media strategy that is much closer to the patient than to their reference institution. Thus, the health information field seems to be distinguished according to three strategic factors: whether or not they are from the health world (32.5% of the sites), whether they are an organization (80% of the sites) or an individual (20%), and whether they belong to the commercial space (45.1% of the sites) or the public space (54.9% of the sites).

---

6 We classify in this category all the actors who propose more or less global therapies based on non-scientific methods. Mainly methods based on the use of natural, magical or esoteric elements.

We then categorized the messages using a text mining method, automatically categorizing the contents according to an n-dimensional vector position. The main theme that forms the core of the representations aims to provide information on the characteristics of the disease and the behaviors to adopt in a therapeutic educational approach (64% of the pages). This main theme is composed of three thematic axes. The first concerns the description of the disease, its symptoms, its origins and its consequences (roughly a third of the pages), as well as accompanying discourse and advice on how to better live with the disease and, more precisely, on the right actions to live with our disease (a second small third of the pages). A second thematic axis is a very emotionally-charged discourse on the factors causing serious attacks and the serious risks for asthmatic children (23% of pages). In addition, a third, more sensationalist axis occupies 10.5% of the positions, with themes as varied as the death of an infant with asthma in a hospital emergency room (a recurrent controversy that occupied 1.4% of the discourses in our corpus) as well as sensational studies on automobile pollution, intestinal bacteria, the presence of pets (dogs in particular) and drug interaction. Finally comes the promotion of spa treatments for asthmatics (2.7%) by companies in the health sector. Thus, nearly 80% of the pages studied are directly linked to a desire to inform the patient in a rather alarmist manner. A tiny minority are sensational pages that try to create a buzz through scoops.

Given the strong correlation of the variables *emmitter_type*, *media_type* and *themes*, we operationalized the notion of *strategic position* by constructing an eponymous variable. The use of a new Hierarchical Ascending Classification (HAC) allowed us to build seven/eight strategic typologies in the corpus studied:

– The most important strategy, which occupies two tenths of the sites, over represents health institutions, public institutions, associations of practitioners and universities, and finally insurance and mutual insurance sites around the themes of therapeutic education and disease prevention policies. Here, it is the institutional word that predominates.

– The second position (two tenths of the sites) is that of the medical press and the specialized press, which primarily use editorial media to address therapeutic education topics, but with a high emotional value, such as serious crises and medical discoveries. This specialized press engages in the therapeutic education process but through the filter of sensational information in a journalistic paradigm.

– We must add to this a position that is very close (competitive?), but more moderate in its treatment. This third position (one tenth of the sites), made up mainly of associations of practitioners, as well as a small part of the specialized professional medical press, endeavors to relay current health information in order to live better, as well as to prevent serious risks and crises. This position, between that of health institutions and that of the medical and specialist press, does not prioritize information through the prism of the event. Many practitioners working in associations offer more general public information related to the concerns of patients and carers, based on the model of the advice they can provide in their practices.

– On the contrary, the fourth position (one tenth of the sites) is the one through which the scandals occur. Composed essentially by the general press using mainly editorial sites, these pages are clearly in search of scandals and accidents and, to a lesser extent, exceptional scientific discoveries which will attract audiences and sell well. Here we find the archetype of the sensationalist press, driven by the search for scoops whose content is widely spread on social networks.

– The fifth position (two tenths of the sites) represents invested individual discourse. Here, we find the patient's discourse alongside the individual practitioner and part of the alternativists (very marginal). They invest heavily in Web 2.0 with a strong preference for blogs and social networks. Their favorite theme remains therapeutic education, a subject on which they challenge health institutions by addressing practices left on the margins of the medical profession. This category constitutes the active avant-garde of Web 2.0, both expert and non-institutional.

– The 6th position (one tenth of the sites) is very close to the previous one and mainly involves patient associations, individual patients, and the specialized press to a lesser extent. But this social group is that of conversations on forums. Here we find the place of discourse that is very much studied in the sociology of the Web, where patients and their carers are looking for practical advice in the daily management of the disease.

– Finally, the 7th and last position (one tenth of the sites) is that of the commercial strategy of the companies which uses a very complete range of media on the basis of e-commerce sites, as well as of portals and directories, and institutional and thematic sites in order to sell products and related care services (creams and oils, thermal cures, etc.). A very complete communication strategy is then deployed on the basis of an institutional strategy of referencing and influence.

Thus, we must note that the actors occupy very distinctive positions, but remain within a traditional communicational order. Alternativists are, quantitatively, a negligible pinch in the discourse and do not appear (in our study limited to the Web seen by Google) likely to question the established order. The wave of patients' and patients' associations' discourse is very much in the minority compared to the presence of health professionals. The latter have been able to come out of their institutional reserve (individually or in association) by adopting the new socio-numerical media (collaborative websites, individual blogs, Facebook pages or groups, etc.). However, the dominant players in health information remain the media industry through the general, specialized and medical press, which occupies the leading position in terms of speaking out in the field of health information.

3.3.2. *From medical legitimacy to editorial power*

To this sociotypical approach, we must add a performance analysis of the discourse. In our research on the influence of discourse on the Internet, we wanted to identify the social resonance of discourse through likes, shares and comments on two networks (Facebook and LinkedIn). In terms of strategies, those developed by the daily and regional press and the specialized press are the only ones that are clearly correlated with success on social networks. Conversely, health officials, practitioner associations, health bloggers and health forums are significantly under-represented in the modalities of great success on Facebook. When we look closely at the factors, one by one, in order to specify what weighs on the success of these strategies, we see that alternative bloggers are the only type of actor – far behind the general press – that can emerge from social networks. It is in fact the editorial strategy that provides the content necessary for social resonance. The difference in resonance between the press and the alternativists is due to the audience base on which the press relies.

However, the success of the alternative bloggers, considering the means they have, is still remarkable. If we add to these points the analysis of the themes addressed, we discover that it is sensational information that has gained significant popularity on social networks. Scandals such as the news of accidental deaths in a hospital are very clearly a good way to ignite social networks. Next in the ranking are articles based on "recent scientific studies" that make "exceptional" discoveries or the discovery of serious risk factors.

It is thus very clearly the strategy of sensationalism of the press and the alternativists that emerge victorious from the social resonance measure. The medical discourse of therapeutic education and advice for a better daily life not only does not interest the press, but above all does not interest social networks.

In our research, we included explicit citation in the body of the text as an indicator of legitimacy. We wanted to know if the nature of the actors played a role in their website's position in the editorial community. By cross-checking *actors* and *bin_indegree* (being cited more than average) at the page level, we find that not only is the correlation significant [X2 (13, N = 1245) = 47.8, p-value < 0.01], but, according to Pearson's residual analysis, only individual practitioners (3.97), patient associations (3.68) and the medical press (2.78) are significantly over-represented in terms of their chances of being cited. On the other hand, no single topic was significantly more likely to be cited than another. The analysis of authority and page rank confirms the role of the medical press, but even more so that of associations of practitioners and health institutions at the heart of the citation network. Thus, the most legitimized actors in terms of citations are clearly those with institutionalized expertise in health but in forms of intermediation that are more accessible than the official discourse. The medical press and associations of health practitioners enjoy a very powerful legitimacy in terms of co-citation in a context where individual initiatives seem to lack credibility. Institutional legitimacy and a therapeutic education strategy seem to be a source of influence with other Web publishers. It is therefore necessary to be in a position of expert mediator in order to play an authoritative role on the Web.

3.3.3. *A need for increased competence in information retrieval in the health sector*

The informational offer shows that it comes from a diversity of actors sometimes difficult to identify by the users, while the development of informational practices and digital skills in health on the Web are far from being acquired. The results of the qualitative survey highlight a certain complexity of online health practices among the elderly population. The latter are particularly entrenched and have visibly increased over the years within our study sample, an evolution that can be considered the consequence of a triple effect.

The advancement of age, which goes hand in hand with the appearance of new symptoms and/or the aggravation of existing symptoms, in a more or less worrying way. The user may then be tempted to carry out a medical (self)consultation via the medium of the Web in order to assess the need to go to the doctor [VIG 17].

Online health uses tend to develop, especially with the appearance of new health platforms or platforms dedicated to teleconsultation. Thus, over the years, the Web tends to be related to the idea of self-evaluation [FOU 01] and taking charge of our health.

The increase in the ease of use of digital platforms by seniors, due to the effect of usage mimicry (copying the uses of relatives and friends) and the intervention of descendants in the digital mediation [LED 12, LEDO 14] of uses and supports.

"The multiplication of Internet sites devoted to health raises the question of the quality of the information disseminated" [BOU 03], all the more so if we take into consideration the logics of crawling, of information classification, of marketing, influenced by the principle of attention economy, in particular in health matters [VIG 18]. Our analysis of the second part of the qualitative survey tends to show that, although health-related uses are more numerous, their effectiveness, quality and safety are no less problematic. Indeed, the groups presenting risky practices in the first part (alternative self-medication, health research without medical skills or knowledge, refusal to use professionals, etc.) do not show an increase in their skills. Beyond this notion of research, it appears that the refusal of chemical medication and of health monitoring by a professional have, in this group of about 10 people, been largely affirmed.

Christiane, for example, who in 2016 refused a vaccine and did not always follow her doctor's prescriptions, has completely stopped consulting health professionals in cases of acute symptoms or seasonal pathology. Since the end of 2018, she has only resorted to online searches, even though she has a high-risk profile and believes her health has in fact deteriorated over two years. She combines a very low health culture with a research methodology based on a certain associative serendipity, with decision-making for herself and for third parties that can lead to dramatic effects. In 2016, she recommended an ointment to her husband, following a search on health forums, the application of which could have had lethal consequences due to drug interactions.

However, since then, her online health research skills do not seem to have evolved and her rejection of institutional discourse has become stronger. For example, her research methodology relies on health information published on forums, without medical terminology, and other sites whose information is not verified. Christiane's case illustrates the fact that evaluating information is "an operation mobilizing both our knowledge, our skills, as well as our personal values, our judgments, our opinions and our tastes. This operation is neither purely technical, nor purely informational, nor is it purely cognitive, but it intertwines all of these dimensions and many more personal ones. […] The evaluation of the credibility of an author, a text or a piece of information was already a challenging task in the printed world, but it has become much more complicated in the digital world" [SER 12]. The complexity, for this type of profile, is based on multiple difficulties. For one, there are gaps in information research (identification of sources on search platforms and verification of the relevance of information). Furthermore, the health culture and the relationship with the medical field lead to a relatively high risk (lack of vocabulary, interpretation of symptoms, presumed credibility of authors without verification, etc.). Finally, the consideration of commercial and attentional logics seems to be weak, users having little awareness of the marketing dimension linked to the results of health research on the Internet.

One of the main difficulties of the users interviewed is based on the "undifferentiation of sources" when searching for health information online, as well as a "general blurring of sources and document status" [SER 12]. Online, it is far less easy for them to sort between deleterious, commercial or safe information. In addition to this, there are the editorial processes, with the necessary distance between the form of the document and its content. A well-written text, using medical terminology, does not guarantee its scientific and medical relevance but plays on the notion of credibility by using emphasis and scientific rhetoric. Here, we see the link between traditional information literacy and info-digital literacy. Research on the Web thus depends on traditional skills and new informational skills, taking into account the specificities of the digital medium, as well as the personality and general culture of the users. However, this intermingling does not seem to be taken into account by the retired users in the sample. "The development of intellectual technologies of representation and access to information is making the phenomenon of serendipity a more important part of research every day" [ERT 03]. Serendipity does not depend so much on the health literacy of users but rather on their digital literacy and its application to informational research. Thus, if users are able to analyze health information, they experience difficulties in finding it on search engines.

If they avoid open forums and prefer institutional sites, they acknowledge having difficulties in identifying the relevance and legitimacy of third-party sources, particularly those of pure-players dedicated to health. "In the digital world, and even more so on the Internet, we are confronted with a mixture of genres and supporting documentation, with the appearance of forms and support documents specific to the Internet, [...] without an equivalent in the printed world" [SER 12]. Online health research thus presupposes a specific culture, which the elderly are not always equally endowed with. Thus, their health research takes the form of informational browsing, in which users consult a relatively wide variety of sources, while calling upon their health literacy to perform informational sorting. To use Ertzscheid and Gallezot's image, users are looking for the needle in the haystack, without presenting a structured research methodology adapted to the digital medium, despite their level of health literacy.

Thus, the research methodology consists of "randomly locating [...] new keywords, new names of people [or Internet sites] that will lead to another line of research and will constitute an answer/solution to the question/problem" [ERT 03], thus resorting to associative serendipity. Information browsing mobilizes little or no health knowledge and skills. Or at least, the medical and media background of this user profile does not seem sufficient to guarantee a legitimate and relevant result. Within the sample, seniors who use associative serendipity have never worked in the medical field and are interested in this field more as a hobby than as an advanced specialization. The impression of great quantities of information – even of "infobesity" – then acts as an element that satiates the informational appetite, without the results and their interpretation being consistent with the user's initial need. "In this process, the information will be primarily 'captured' by the user and will be used as a basis for the associations that he will build in order to complete his quest" [ERT 03]. However, what good is such an association if health literacy does not allow for a justified and relevant analysis of the information – information which itself tends not to be based on a structured approach that is adapted to the challenges of attention capture?

Moreover, the risk of such health uses, combined with the "serendipitous" nature of online searches, relies on the subject's ability to see themselves as an object. "By becoming a doctor for oneself, we must study the constitution of the subject as an object for itself, the disease becoming an object of verification and control of oneself by oneself" [AND 10]. Searching for health information online, for ourselves or for others, presupposes a necessary distancing and stepping back on the part of the user, so as not to fall into emotion or subjectivity. It is therefore essential that this same user has the skills and knowledge to enable them to objectify the results of their research. This applies as much to the informational dimension of research (relevance of the sources used, etc.) as to the health dimension (identifying the disease linked to the symptom or pathology) and to the media dimension (getting rid of the commercial aspect of certain information).

3.4. Health, information and digital technology: a logic in line with transliteracy

Between understanding the informational offer on the Internet and analyzing patients' health literacy, our research highlights the need to use the notion of transliteracy [THO 07, LED 14] in the analysis of health information on the Internet. Literacies depend on an alliance between:

– information literacy: searching, finding, analyzing and reusing information, etc.;

– digital literacy: using a search engine, knowing how to classify information, detecting fraudulent and commercial information, identifying relevant sources to verify, etc.;

– health literacy: identifying a symptom, handling a dosage, checking for contraindications and drug interactions, etc.

As the informational and technical context has evolved, health literacy cannot remain the same as it was at the beginning. In this respect, it is impacted in the same way as other information and media literacies, under the effect of the evolution of the information on offer. In this context, the user or patient becomes a co-producer of potential information. From the opinion mentioned on a platform, to active participation in communities of patients with chronic diseases, the place of users in health information processes has quite clearly evolved. Nevertheless, this communicative logic shows differences in users' mastery and skills. These differences are in addition to the initial difficulties of health literacy, with regrettable skill deficits in the ability to search for information, as well as difficulties in understanding dosage and its application.

3.4.1. *An extension of competences to be put into context*

This evolution makes the analysis of users' skills in the field more complex, as the evaluation of an individual's abilities cannot be based on the same criteria as in the 1980s. In this respect, health literacy is met by the same concerns as media literacy and information literacy, which have been strongly impacted by the development of new media, by the potential for users to produce information and by access to new types of information on a variety of media. Consequently, health literacy cannot be understood without digital media. This is why we have proposed to redefine it as digital health literacy, as computerized or digitized processes interfere with medical activities and patients' practices.

This choice has also been expressed at the level of the European Union, which has funded projects dedicated to digital health literacy in order to improve it among European citizens. One of the avenues explored is producing online MOOCs in order to develop the empowerment of European citizens[7]. If the diagnosis seems relevant, the MOOC approach seems to be a questionable interpretation, as the skills mentioned and described above depend on a strong economic, social and cultural context. It is the first stumbling block in a complex literacy, which mixes informational and digital issues with health skills, that we are tempted to neglect the context in an approach that would favor entry by a series of skills.

However, the literature on health literacy has always emphasized the importance of context. Research has often focused on socially disadvantaged populations to better understand where the key issues lie. In this context, the research shows that there are often great difficulties in understanding the organization of the health system and that vocabulary difficulties far outweigh the desire to make requests on a search engine. Indeed, the specific vocabulary of health professionals, coupled with those of medical services,

7 This is the case of the IC-Health project (Improving Digital health literacy in Europe), whose real contributions are slow to be seen. https://ichealth.eu/.

is sometimes difficult to understand for patients, especially since the public may be allophone or even illiterate.

Several studies have examined methods of assessing health literacy [BAK 99, PAR 95]. Various tools have been developed for this purpose, in particular to highlight training gaps and deficits that are more or less worrying depending on the populations studied. These tools include the Rapid Estimate of Adult Literacy in Medicine (REALM) and the Test of Functional Health Literacy in Adults (TOFHLA). Assessment methods are often based on questionnaires, which can be used at the national level. It seems appropriate to focus on the Health Literacy Questionnaire (HLQ), developed in Australia, which has been adapted internationally. The questionnaire allows a variety of issues to be addressed. Nine of them are highlighted:

- feeling understood and supported by healthcare providers;
- having sufficient information to manage my health;
- actively managing my health;
- social support for health;
- appraisal of health information;
- ability to actively engage with healthcare providers;
- navigating the healthcare system;
- ability to find good health information;
- understanding health information well enough to know what to do.

The objective of the questionnaire is to be able to identify avenues of support and remediation for users. With several colleagues [BAT 14], Richard Osborne has developed a dedicated approach for this purpose: OPtimising HEalth LIteracy and Access (OPHELIA).

3.4.2. *A transliteracy to be modeled*

Consequently, digital health literacy is in line with existing observations regarding the need for training in digital and information literacy. It cannot really properly develop without basic literacy and social conditions that truly allow the user to make informed choices with a free conscience. To cope with the impression of an elite literacy, which intersects hybrid competences, it is possible to consider the path of transliteracy as opportune. According to the definition of [THO 07]. "Transliteracy is the ability to read, write and interact across a range of platforms, tools and media from signing and orality through handwriting, print, TV, radio and film, to digital social networks".

Transliteracy should therefore be understood as cumulative literacy, building on the knowledge and skills that users have acquired in other circumstances and that can be reinvested. This aspect appears essential as a response to the idea of producing MOOCs that could be addressed to all types of audiences. This is the second lesson. It is not possible to hope to increase the level of digital health literacy without having educational systems that provide an excellent level of information and media literacy, as well as a good level of digital literacy. At the same time, quality training in health information is essential to ensure that people do not have difficulty understanding not only the healthcare system, but also basic anatomical knowledge, without which self-awareness is severely lacking.

3.4.3. *Data literacy: a new perspective*

A fourth sector is now developing with regard to digital health literacy: the issue of health data. There is a potential reversal of perspective with the fact that the production of health information by users and patients has become a coveted object that adds to an informational offer on the Web, already highly prized by commercial sites. Connected objects and associated services allow the production of personal health data in addition to those already produced by official services. This personal health data includes data contained in medical records, such as the shared medical record (DMP), as well as data retrieved by connected devices prescribed or used autonomously, to which can be added all the data from searches for information on search engines, consultation histories, as well as consultations and registrations on forums that deal with health information.

While the summary typology outlined here gives the impression of separate silos, doubts may nevertheless be raised, including in Europe, where the stakes of takeover and recovery are of concern, especially since they involve a series of actors who have been somewhat on the sidelines until now[8]. Subsequent reuse, unforeseen or insufficiently described, as well as contracts for use, leave the door open to potential decisions that could be taken by insurance companies and mutual insurance companies or other actors. Even the organizations that make it possible to establish DNA and possible ancestry, which target their advertising on the question of origins, are nevertheless building up databases with considerable medical potential.

The current work on data literacy [KOL 15] raises the issue of training for leading-edge workers to meet the demands of a world of work, and seeks to identify ways to take a citizen view of the data made available by open data. As a result, the issue of personal health data raises issues of paramount importance that cannot be resolved by a few legislative safeguards and a simple desire for empowerment [LAB 19]. Ultimately, the data component of digital health literacy could be the element that highlights the strong inequality that exists between citizens, with mechanisms that could lead to a divide between those able to retain forms of control over their data and those forced to make it available to third parties in order to benefit from seemingly less costly services.

3.5. Conclusion

The first phase of the study highlighted several original findings. This quantitative approach aimed to map and analyze the architecture of health information available on the Web through the example of asthma in children. The strategic positions of the influencers are sufficiently typified to be able to speak of a digital microcosm. The press and health professionals remain the dominant players in health information on the Internet, with more than 70% of the pages produced. Opposite them, the patient's voice is in the minority (14%). Even if the press is constantly looking for sensational information, novelties and unusual facts following a logic of "breaking news" and ratings, there is both a specialization and a hierarchization in the strategies of the actors that emanate from the editorialization of sensational and exceptional information to therapeutic education, mediated by health professionals. The core of the network in terms of influence remains a hierarchical three-way game between the medical press, practitioners'

8 One example is the fact that, in the UK, Boris Johnson's government has granted Amazon access to medical information stored on NHS (National Health Service) servers.

associations and health institutions. The discussion of the intimate and daily concerns of the patient and the commercial aspect of health remain very weak in the digital public space studied.

If we add the very weak presence of the alternativists (less than 3% of the pages accessible in the search engines) to the corpus, we have to admit that the social order of health information discourse is challenged very little in the digital era, both in terms of quantity (content, sites, etc.) and in terms of Web influence. However, we feel the presence of highly specialized information content in the context of discussions on daily life associated with Web 2.0. As mentioned, the need for research on the mediated Web within social networks appears essential to properly measure this phenomenon.

The constantly debated definition of digital health literacy shows the difficulty in giving a good assessment of it. There are a few health literacy indicators that now incorporate informational and digital issues [OSB 13]. While they may be interesting and timely at a given moment, the indicators require updating, which ultimately makes it difficult to carry out long-term monitoring, or even to claim to obtain improved results since the territory of this literacy continues to become more complex as time goes by. It is therefore impossible to think of it too separately from the other literacies. A few years ago, we tried to produce a digital health literacy indicator with an online test[9]. This tool mixed basic health and dosage questions with informational and digital questions. It should be revised by integrating the relationship with connected objects and data. The results of the test showed a considerable deficit in the understanding of the functioning of search engines, particularly concerning methods for ranking results. This lack of algorithmic understanding already showed the need for search engine literacy [LED 16, LED 17], which is now coupled with the issues of personal data and possible automated processing. This complexity raises questions about training issues as well as about survey and analysis methods, which are forcing public health fields to be increasingly open to hybrid methods and transdisciplinary associations. Presumably, this also leaves the door open to the possibility of anthropological work, which integrates these different findings to propose services that do not leave the responsibility of training solely to users and patients.

---

9 Digital Health Literacy Placement Test: http://www.megatopie.info/testor/enquete.html?e=19.

Nevertheless, even in this framework, as a complement to this study, we would have to conduct comparative studies on the corpora accessible on the Web (including social networks) against the browsing data of a cohort of users, to be able to consolidate our results by comparison. But only future developments of our project (allowing a close crawl of relevant documents with respect to a dictionary *a priori*) would allow us to introduce a real snowball sampling. It is on this condition that we could free ourselves from the framing wanted by the search engines to draw a more representative Web architecture. We could then study the effect of this architecture as public health factors influencing the health literacy of a cohort of patients.

Questioning health literacy challenged by digital technology forces us to rethink training strategies in this area, in order to respond to increasingly important issues at both the individual and collective levels. The identification of competencies certainly appears useful, but history shows that this hybrid literacy clearly refers to elitist levels of competence, if we seek to reach a minimum level. The will to make the citizen become more competent seems commendable, provided that training and its articulation are rethought. But it cannot be enough, as the conditions in which it is exercised vary from reading a dosage to querying a search engine, via connected devices. Above all, it seems appropriate to advocate for a strengthening of mediation mechanisms. In the short term, this implies better training for healthcare professionals in these issues, as well as the recruitment of intermediate professionals capable of performing this mediation. Indeed, it should not be forgotten that this is not only a series of skills to be mobilized and exercised but also a form of social, cultural and civic mediation.

3.6. References

[AND 10] ANDRIEU B., *Le monde corporel : de la constitution interactive du soi*, L'Âge d'Homme, Paris, 2010.

[BAK 99] BAKER, D.-W., WILLIAMS M.-V., PARKER R.-M. *et al.*, "Development of a brief test to measure functional health literacy", *Patient Education and Counseling*, vol. 1, no. 38, pp. 33–42, 1999.

[BAR 05] BARABASI A.-L., "The architecture of complexity: The structure and the dynamics of networks, from the Web to the Cell", *Proceedings of the Eleventh ACM SIGKDD International Conference on Knowledge Discovery in Data Mining*, Association for Computing Machinery, New York, available at: https://doi.org/10.1145/1081870.1081873, 2005.

[BAR 12] BARBILLON E., LE ROY J., *Petit manuel méthodologique de l'entretien de recherche : de la problématique à l'analyse*, Enrick B. Editions, Paris, 2012.

[BAR 14] BARABASI A., *Linked: How Everything is Connected to Everything Else and What It Means for Business, Science, and Everyday Life*, Basic Books, New York, 2014.

[BAR 17] BARATS C., *Manuel d'analyse du web*, Armand Colin, Paris, 2017.

[BAT 13] BATTAÏA C., L'émotion dans les forums de discussion : structuration et évaluation de l'information de santé, available at: http://www.theses.fr/2013GRENL012, 2013.

[BAT 14] BATTERHAM R.-W., "The OPtimising HEalth LIterAcy (Ophelia) process: Study protocol for using health literacy profiling and community engagement to create and implement health reform", *BMC Public Health*, vol. 1, no. 114, pp. 680–694, 2014.

[BAW 01] BAWDEN D., "Information and digital literacies: A review of concepts", *Journal of Documentation*, vol. 2, no. 57, pp. 218–259, 2001.

[BLO 08] BLONDEL V.-D., GUILLAUME J.-L., LAMBIOTTE R. *et al.* "Fast unfolding of communities in large networks", *Journal of Statistical Mechanics: Theory and Experiment*, vol. 10, 10008, 2008.

[BOU 03] BOURRET C., "La santé en réseaux", *Études*, vol. 9, no. 399, pp. 175–189, 2003.

[BOU 16] BOULLIER D., *Sociologie du numérique*, Armand Colin, Paris, 2016.

[BRA 05] BRANDES U., ERLEBACH T., *Network Analysis: Methodological Foundations*, Springer, Geneva, 2005.

[BRA 10] BRANDES U., "A faster algorithm for betweenness centrality". *The Journal of Mathematical Sociology*, vol. 2, no. 25, pp. 163–177, 2010.

[BRI 05] BRIN S., PAGE L., "The anatomy of a large-scale hypertextual Web search engine", *Computer Networks and ISDN Systems*, vols 1–7, no. 35, pp. 107–117, 2005.

[CON 13] CONVERT B., DEMAILLY L., "Les interactions dans Internet", *Communication – Information médias théories pratiques*, vol. 22, no. 2, pp. 63–98, 2013.

[COR 19] CORMERAIS F., LAKEL A., "Recherches digitales et production des données, bouleversement des agencements pour le chercheur en SIC", *Etudes Digitales*, vol. 3, no. 6, pp. 94–111, 2019.

[DEB 18] DEBUSSCHE X., LENCLUME V., BALCOU-DEBUSSCHE M. *et al.*, "Characterisation of health literacy strengths and weaknesses among people at metabolic and cardiovascular risk: Validity testing of the health literacy questionnaire", *SAGE Open Medicine*, available at: https://doi.org/10.1177/2050312118801250, 2018.

[ERT 03] ERTZSCHEID O., GALLEZOT G., "Chercher faux et trouver juste, sérendipité et recherche d'information", *Colloque bilatéral Franco-Roumain, CIFSIC Université de Bucarest*, available at: https://archivesic.ccsd.cnrs.fr/sic_00000689/document, 2003.

[ESP 13] ESPITALIER F., BARBAZ M., FUSCIARDI J. *et al.*, "Péridurale obstétricale : qualité de l'information médicale en français sur Internet en 2009 et 2010", *Annales françaises d'anesthésie et de réanimation*, vol. 32, no. 9, pp. 554–559, 2013.

[ESP 14] ESPITALIER F., DE LAMER S., BARBAZ M. *et al.*, "Qualité de l'information médicale et péridurale obstétricale : quoi de neuf en 2013 ?", *Annales françaises d'anesthésie et de réanimation*, vol. 33, no. 2, pp. 351–352, 2014.

[FRÉ 01] FRÉMONT P., LABRECQUE M., LÉGARÉ F. *et al.*, "Evaluation des sites Web médicaux : fidélité interobservateur et intraobservateur d'un outil d'évaluation", *Canadian Family Physician*, vol. 47, pp. 2270–2278, 2001.

[FOU 01] FOUCAULT M., *Dits et écrits, 1954–1988. I, 1954–1975*, Gallimard, Paris, 2001.

[GRA 87] GRAFF H.J., *The Legacies of Literacy: Continuities and Contradictions in Western Culture and Society*, Indiana University Press, Delhi, 1987.

[JEN 16] JENDOUBI S., Influencers characterization in a social network for viral marketing perspectives, PhD thesis, University of Rennes 1, available at: https://tel.archives-ouvertes.fr/tel-01665815/document, 2016.

[KLE 99] KLEINBERG J.M., "Authoritative sources in a hyperlinked environment", *Journal of the ACM*, vol. 46, no. 5, pp. 604–632, available at: https://doi.org/10.1145/324133.324140, 1999.

[KOL 15] KOLTAY T., "Data literacy: In search of a name and identity", *Journal of Documentation*, vol. 71, no. 2, pp. 401–415, available at: https://doi.org/10.1108/JD-02-2014-0026, 2015.

[LAB 19] LABARTHE F., TRAINOIR M., "Explorer les déclinaisons de l'empowerment en régime numérique", *Terminal*, nos 125–126, available at: https://doi.org/10.4000/terminal.5570, 2019.

[LAK 08] LAKEL A., "Penser la gouvernance de l'Internet. Le pouvoir dans la recherche sur les TIC", *16ème congrès 2008 de la SFSIC : épistémologie des SIC : émergences, opportunités, affirmations*, Compiègne, April 2008.

[LAK 16] LAKEL A., LE DEUFF O., "À quoi peut bien servir l'analyse du Web ?", *Les Cahiers du numérique*, vol. 13, no. 3, pp. 39–62, 2016.

[LAK 19] LAKEL A., "Prises de positions et influences sur le Web : le cas de l'information de santé", *Revue française des sciences de l'information et de la communication*, vol. 18, available at: https://doi.org/10.4000/rfsic.8376, 2019.

[LAT 07] LATOUR B., *Reassembling the Social: An Introduction to Actor-Network-Theory*, 1st edition, Oxford University Press, Oxford, 2007.

[LAT 10] LATOUR B., "Networks, societies, spheres: Reflections of an actor-network theorist", *International Seminar on Network Theory: Network Multidimensionality in the Digital Age*, Annenberg School for Communication and Journalism, Los Angeles, available at: ffhal-00972865, February 2010.

[LAT 14] LATOUR B., *Changer de société, refaire de la sociologie*, La Découverte, Paris, 2014.

[LED 11] LE DEUFF O., *La formation aux cultures numériques : une nouvelle pédagogie pour une culture de l'information à l'heure du numérique*, FYP, Limoges, 2011.

[LED 12] LE DOUARIN L., CARADEC V., "TIC, grands-parents et leurs petits enfants : vers un conflit des générations technologiques ?", in CHAMAHIAN A., LEFRANÇOIS C. (eds), *Vivre les âges de la vie*, L'Harmattan, Paris, 2012.

[LED 14a] LE DEUFF O., "Translittératie et transmédia : quelles compétences pour de nouvelles productions 'délivrées' ?", *Les Cahiers du numérique*, vol. 10, no. 3, pp. 55–72, 2014.

[LED 14b] LE DOUARIN L., "Usages des nouvelles technologies en famille. Informations sociales", vol. 181, pp. 62–71, available at: http://www.cairn.info/revue-informations-sociales-2014-1-p-62.htm, 2014.

[LED 16] LE DEUFF O., "La littératie digitale de santé, un domaine en émergence", in *Les écosystèmes numériques : intelligence collective, développement durable, interculturalité, transfert de connaissances*, AGOSTINELLI S., KOULAYAN, N., (eds). Presses des Mines, Paris, 2016.

[LED 17] LE DEUFF O., "Search engine literacy. Information literacy in the workplace", in KURBANOĞLU S., BOUSTANY, J., ŠPIRANEC, S. *et al.* (eds), Springer, Geneva, 2017.

[MON 08] MONNOYER-SMITH L. "Pour une épistémologie complexe des SIC", *Actes du 16e congrès de la SFSIC*, available at: https://www.sfsic.org/congres_2008/spip.php?article78, 2008.

[OSB 13] OSBORNE R.H., "The grounded psychometric development and initial validation of the health literacy questionnaire (HLQ)", *BMC Public Health*, vol. 13, no. 1, pp. 658–675, 2013.

[PAG 14] PAGANELLI C., CLAVIER V., DUETTO A., "S'informer via des médias sociaux de santé : quelle place pour les experts ?", *Le Temps des médias*, vol. 23, pp. 141–143, 2014.

[PAI 12] PAILLÉ P., MUCCHIELLI A., *L'analyse qualitative en sciences humaines et sociales*, Armand Colin, Paris, 2012.

[PAR 95] PARKER R.-M., BAKER D.W., WILLIAMS M.-V. *et al.*, "The test of functional health literacy in adults", *Journal of General Internal Medicine*, vol. 10, no. 10, pp. 537–541, 1995.

[PEE 09] PEERSON A., SAUNDERS M., "Health literacy revisited: What do we mean and why does it matter?", *Health Promotion International*, vol. 24, no. 3, pp. 285–296, available at: http://doi.org/10.1093/heapro/dap014, 2009.

[PRO 18] PROULX S., RUEFF J., *Actualité des méthodes de recherche en sciences sociales sur les pratiques informationnelles*, Centre d'études sur les médias, available at: https://www.cem.ulaval.ca/wp-content/uploads/2018/12/methodespratiques.pdf, 2018.

[ROS 17] ROSSI M., Graph mining for influence maximization in social networks, PhD thesis, Paris-Saclay University, Paris, available at: https://pastel.archives-ouvertes.fr/tel-01688005/document, 2017.

[SCO 17] SCOTT J., *Social Network Analysis*, 4th edition, SAGE Publications Inc, Thousand Oaks, 2017.

[SER 12] SERRES A., *Dans le labyrinthe : évaluer l'information sur internet*, C&F éditions, Paris, 2012.

[SIM 74] SIMONDS S.K., "Health education as social policy", *Ikaith Education Monograph*, vol. 2, pp. 1–25, 1974.

[STA 10] STAII A., PAGANELLI C., ROMEYER H. *et al.*, Dynamiques interactionnelles et rapports à l'information dans les forums de discussion médical, available at: http://hal.univ-grenoble-alpes.fr/hal-01135102, 2010.

[THO 07] THOMAS S., JOSEPH C., LACCETTI J. *et al.*, "Crossing divides", *First Monday*, vol. 12, no. 12, pp. 3–12, 2007.

[VIG 15] VIGOUROUX-ZUGASTI E., LE DEUFF O., "Le senior connecté", in SALEH I., CARAYOL V., LELEU-MERVIEL S. *et al.* (eds), *H2PTM'15 : Le numérique à l'ère de l'Internet des objets, de l'hypertexte à l'hyper-objet*, ISTE Editions, London, available at: https://archivesic.ccsd.cnrs.fr/sic_01272666/document, 2015.

[VIG 16] VIGOUROUX-ZUGASTI E., LE DEUFF O., Rapport final du projet Risque Informationnel Chez les Seniors et Automédication (RICSA), Bordeaux Montaigne University, Pessac, available at: https://www.fondation-maif.fr/up/pj/rapportfinalricsa.pdf, 2016.

[VIG 17] VIGOUROUX-ZUGASTI E., L'expérience du numérique chez les aînés : enjeux et limites du solutionnisme technologique, PhD thesis, Bordeaux Montaigne University, Pessac, available at: https://tel.archives-ouvertes.fr/tel-01764923/document, 2017.

[VIG 18] VIGOUROUX-ZUGASTI E., "Experts et dispensateurs de conseils de santé en ligne : (en)jeux d'attention et de crédibilité", *Communication & Organisation*, vol. 53, pp. 45–158, 2018.

[WAS 94] WASSERMAN S., FAUST K., *Social Network Analysis: Methods and Applications*, 1st edition, Cambridge University Press, Cambridge, 1994.

[ZUR 74] ZURKOWSKI P.G., The information service environment: Relationships and priorities, National Commission on Libraries and Information Science, Washington, DC 1974, Report ED 100391, available at: https://files.eric.ed.gov/fulltext/ED100391.pdf, 1974.

4

The Essential Role of Health Information Literacy in Seniors' Healthcare Communication

4.1. Introduction

Western societies are aging, with the share of people aged 60 years and over growing very fast, generally because life expectancy is increasing. This will have substantial consequences across social, political and socioeconomic levels (Wilkowska *et al.* 2018). The public sector, not least the healthcare sector, will be affected by this, because being older increases a risk of a deterioration in health (e.g. Berkowsky and Czaja 2018). The healthcare services facing an increase in the number of potential patients need to offer high-quality communication, as the quality of communication is related to proper care and adherence to care and medications (e.g. Parker *et al.* 2020) as well as to patient satisfaction (e.g. Trummer *et al.* 2006). Older adults in particular may experience greater difficulties in accessing or comprehending health-related information (e.g. Tennant *et al.* 2015). Older age increases the prevalence of cognitive impairment and memory problems, as well as chronic diseases. Both physical and mental health may deteriorate, and sensory impairments, such as a deterioration in vision and hearing, become more prevalent (Sanchiz *et al.* 2017; Berkowsky and Czaja 2018). These changes are bound to influence the understanding of health-related information. Furthermore, older adults may have difficulties with processing and remembering information (Posma *et al.* 2009; Sanchiz *et al.* 2017).

Chapter written by Kristina ERIKSSON-BACKA and Stefan EK.

Gender	
Female	**57% (159)**
Male	**43% (122)**
Total	***100% (281)***
Age group	
65–69 years	49% (137)
70–74 years	32% (90)
75–79 years	19% (54)
Total	*100% (281)*
Education level	
Basic (comprehensive school education)	34% (95)
Intermediate (upper secondary or vocational school education)	44% (124)
Higher (degree from university or university of applied sciences)	22% (60)
Missing	1% (2)
Total	*100% (281)*

Table 4.1. *Socio-demographic background of the 281 respondents*

Older people clearly have their own health issues, of which interacting with health information both in their everyday life and in healthcare contexts is a major one. This chapter will discuss health information behavior in older people, that is, their interest in, seeking activity, and the sources used for health information, as well as their health information literacy, what it is, how it can be measured, and its relation to the more widely used and broader concept of health literacy. Empirical results on how older Finnish people experience communication in healthcare contexts and what kind of role health information literacy can play concerning perceptions of patient–practitioner communication quality are presented. The empirical data for the study was gathered through a survey sent by post to a population of 1,000 Finns aged 65–79 years, who were randomly selected from the Finnish Population Information System, the national computerized system containing basic information on all Finnish citizens as well as foreign nationals residing in Finland. In addition to governmental bodies, businesses and private organizations can also gain access to information such as name, address and year of birth (Digital and Population Data Services Agency n.d.). A total of 281 completed responses were returned. The socio-demographic background information of the respondents is shown in Table 4.1. The majority of the respondents were female (57%), nearly half of them belonged to the youngest age group (65–69 years), and an intermediate level of education was the most common (44%).

4.2. Health information behavior of older adults

According to a well-established definition, information behavior is "the totality of human behavior in relation to sources and channels of information, including both active and passive information seeking, and information use" (Wilson 2000, p. 49). A general assumption is that due to a higher likelihood of decreased health status, the need for information related to health and illness increases with higher age (e.g. Morey 2007; Tennant *et al.* 2015). Older adults indeed seem to be quite interested in obtaining health information. In a previous survey of a small group of elderly Finns, aged 65 or over, 28% were very interested in information about health or illness, and 54% were fairly interested. In addition, 20% of the respondents often sought health-related information and 47% did it fairly often (Eriksson-Backa 2008; Eriksson-Backa 2010). Another Finnish survey of people of retirement age showed respondents that were even more interested. As many as 41% were very interested and 52% quite interested in information related to health (Eriksson-Backa 2011).

The questionnaire employed in this study also contained questions on interest in and seeking activity of health-related information, as well as lists of 13 named health information sources, and the respondents were asked to rate them on a five-point scale regarding how much they were used ("much" to "not at all") and how trustworthy they were perceived ("very" to "not at all"). In addition, the option "do not know" was available. The sources were either traditional media sources, interpersonal sources or electronic resources. The results show that the respondents were generally interested in information about health or wellbeing, as 73% (204/280) had ticked that they were very or fairly interested. Only 7% (20/280) were only a little or not at all interested in such information. Slightly fewer were active seekers of such information, however, as 62% (173/278) answered that they fairly or very often seek such information. Fifteen percent (42/278) were only a little or not at all active information seekers.

The preferred sources were mainly medical. Information attached to medical packages, which could include prescriptions and patient information leaflets, was the source the respondents used the most: 44% (124/278) had replied that they use them often, and 29% (81/278) had ticked that they use them fairly often. The written sources were followed by staff in pharmacies, as 26% (72/278) used these often as sources and an additional 34% (94/278) fairly often. Pharmacy staff were closely followed by healthcare professionals who were often used by 25% (69/277) of the respondents and fairly often by 35% (98/277). These were followed by daily news media with newspapers and magazines in fourth place (17%, 46/277 used them often and 27%, 75/277 fairly often) and TV and radio in fifth place (14%, 37/274 used them often and 29%, 79/274 fairly often). Internet sources came in sixth place with 12% (32/268) using them often and 14% (38/268) using them fairly often.

The most used information sources were considered the most trustworthy, as well, although pharmacy staff came in first place ahead of medical packages. Staff at pharmacies were considered very trustworthy by 42% (117/279) and fairly trustworthy by 44% (122/279). Medical packages were ticked as very trustworthy by 40% (110/277) of the respondents, and fairly trustworthy by 43% (120/277). Slightly fewer, 38% (104/277), considered healthcare professionals very trustworthy, and 49% (136/277) thought that they were fairly trustworthy. Scientific journals, which were often used by only 6% (17/272), came in fourth place regarding trustworthiness (15%, 43/276 considered them very trustworthy and 37%, 104/276 fairly trustworthy), followed by reference books, that were much used by a mere 7% (20/271) but considered very trustworthy by 14% (39/276) and fairly trustworthy by 40% (111/276). Internet resources were down in 9th place, overtaken by family, TV/radio and leaflets. Four percent (11/264) considered Internet resources very trustworthy, and 17% (44/264) thought that they were fairly trustworthy.

The results, hence, largely remind us of the previous ones showing that older people often prefer purely medical sources for health-related information. An earlier small-scale survey of elderly Finns also showed that medical packages together with physicians were the most used sources, closely followed by newspapers and pharmacy personnel. Physicians were, however, most often considered very trustworthy, followed by medical packages, other health professionals such as nurses, and specialists like dieticians or physiotherapists. Physicians and medical packages were also most often considered very important sources (Eriksson-Backa 2008; Eriksson-Backa 2010). Similar results have been found among older Canadian women, who primarily consulted physicians as sources for health information (60%). In contrast to the respondents of this study, they did, however, put Internet sources in second place (59%). These were followed by pharmacy and hospital staff with considerably lower shares (24% and 18%, respectively) (Harris *et al.* 2006). Morey (2007) found that in an African-American community, health service professionals were by far the most important sources and used by a majority of those aged 65–74 years. Medlock *et al.* (2015) presented similar results, as face-to-face contact with health professionals, closely followed by the Internet and pharmacists were most commonly used by older Dutch people, followed by leaflets, newspapers and health magazines. Health professionals and pharmacists were, however, considered very much trusted by around 80%, whereas the Internet was trusted a lot by 42%. In addition, Hall *et al.* (2015), in a study of U.S. citizens aged 50–92 years, also found that health professionals constituted the most preferred sources (76%). Other sources such as Internet sites, magazines, brochures and books or family and friends were mentioned considerably more seldom.

As shown above, this study proves a striking gap between the number of respondents who perceived health professionals and information attached to medications as very important and trustworthy, and the number who thought so about other kinds of sources. The reliance on medical expertise as sources points to the importance of appropriate information provision and communication between health practitioners and older patients. Concerning the patients, the reception of and acting on given information is largely dependent on their skills and competencies in managing the information, which are also called literacies.

4.3. Literacies related to health

People's ability to obtain and make use of health information is important for their ability to take care of themselves and their health. Health literacy is considered one of the crucial prerequisites for successful communication within healthcare. The concept includes the ability to obtain, process, understand and use health information (Rudd 2015). The concept of health information literacy was generated and defined by combining health literacy with information literacy (Shipman *et al.* 2009). The abilities considered important for an information literate person are: the ability to determine the extent of information needed; to access the needed information effectively and efficiently; to evaluate information and its sources critically; to incorporate selected information into one's knowledge base; to use information effectively to accomplish a specific purpose; and to understand the economic, legal and social issues surrounding the use of information, and access and use information ethically and legally (American Library Association 2000, pp. 2–3).

The combined concept, consequently, covers the abilities to recognize a need for health information, to know how and where to find information about health, and how to evaluate and use this information in everyday life to make good health decisions (Shipman *et al.* 2009). The novel concept of health information literacy was previously mainly deployed in either library or educational contexts (e.g. Ivanitskaya *et al.* 2006; Kloda 2008; Li and Brilla 2008). More recently it has, however, in line with an increased change of focus of information literacy research overall, also been used in everyday life settings (Martzokou and Sayyad Abdi 2017).

In addition to these two concepts, Norman and Skinner (2006) coined the concept eHealth literacy, which has been assessed by the eHealth Literacy Scale, eHEALS, that includes eight statements to be answered on a five-point Likert-type scale. It measures how knowledgeable and comfortable people think they are at finding, evaluating and applying online health information, as well as their perceptions of their own skills (Norman and Skinner 2006). In brief, eHealth literacy calls for "a mix of health, information, scientific, media, computer, and Internet literacy" (Watkins and Xie 2014).

4.4. Literacies and older age

An abundance of studies on health literacy have over the past decades used the Test of Functional Health Literacy in Adults (TOFHLA) (Parker *et al.* 1995) or its short-form S-TOFHLA, to study health literacy levels and factors behind differences in health literacy. Age has often proved to be significant, especially being 65 years or older, which has been related to a lower average level of health literacy (Ratzan and Parker 2006). Baker and colleagues (Baker *et al.* 2000, 2002) found strong links between inadequate levels of functional health literacy, "the ability to read and understand health-related materials" (Baker *et al.* 2002, p. 1278) and higher age, as well, which was thought to depend on changes in cognitive functions and a decreased reading ability (Baker *et al.* 2000). Serper *et al.* (2014) found that cognitive function significantly explained relationships between higher health literacy and better health status in older adults. In addition, Baron-Epel *et al.* (2007) found relationships between lower scores and both higher age and lower levels of education.

Another tool, the Health Literacy Questionnaire (HLQ) (Osborne *et al.* 2013), is multidimensional and comprises nine independent scales. Among Australian adults, being 65 years or older and having lower education levels were the strongest reasons for lower scores in some of the scales (Jessup *et al.* 2017). A Danish survey that used some dimensions of the HLQ likewise found that a lower educational level was associated with lower health literacy, as was age when it came to those aged 85 years or older. For those between 65–84 years, problems occurred with reading and understanding information, but they had fewer difficulties in communicating with healthcare providers (Bo *et al.* 2014).

In Europe, use of the evermore common European Health Literacy Survey Questionnaire with 47 items (HLS-EU-Q47) (Sørensen *et al.* 2013) showed lower levels of health literacy for older people (65 years or over) than for younger age groups (Palumbo *et al.* 2016; Schaeffer *et al.* 2017). The short version HLS-EU-Q16 used in studies on older adults show partly similar results: among Germans aged 55-91 years, 23% had excellent, 50% sufficient, 23% problematic and 4% inadequate levels of health literacy (Tiller *et al.* 2015). Preliminary results of Finns aged 75 years showed that 12% had excellent levels of health literacy, whereas 51% had adequate levels, 32% had problematic levels, and 5% insufficient levels (Eronen 2018).

Tennant *et al.* (2015) used the eHEALS scale (Norman and Skinner 2006) in their study of people aged 50 or older and found that eHealth literacy scores decreased with increased age, and that higher levels of education were related to higher scores, that is, not only health literacy but also eHealth literacy is related to age and education level. Neter and Brainin (2012) found similar results in an adult Israeli population.

Also health information literacy has to some extent been studied in relation to older age, although not nearly as extensively as health literacy. Eriksson-Backa (2010) conducted a small-scale survey on Swedish-speaking Finns aged 65 years or older and found: (1) that they thought it was easy to recognize needs for health information and to use it to, for example, follow advice; (2) that they were fairly active health information seekers; (3) that they preferred only a few sources; (4) that they had fairly weak abilities in assessing the quality and (5) to some extent problems with understanding obtained health information. In a review article, Suri and colleagues (2014) identified three research strands connected to health information literacy in literature from 2004 to 2014. Studies have looked at the influence of demographic factors on health information seeking and use; how senior citizens comprehend health information and what impact e-health literacy interventions have. In another review, Hallows (2013) found conflicting results on whether the Internet has influenced the health information literacy of elderly people.

Enwald *et al.* (2018) examined the everyday health information literacy of Finns aged 65 or older and found that around 60% had trouble with knowing who to believe in health issues and how to understand terminology. Huvila and colleagues (Huvila *et al.* 2019) studied differences in health information literacy between people from different age groups in Sweden, among younger adults (born after 1960), older adults (born 1946–1960) and elderly (born 1945 or earlier). They found that the older age groups were less inclined to evaluate health information and more prone to experience information overload. The oldest respondents also had trouble in understanding medical package labels as well as in identifying information needs. Yang (2019), furthermore, surveyed Chinese 50–70-year-olds and found low levels of health information literacy.

4.5. Health literacy and healthcare communication

Health literacy is important because research has linked limited health literacy to poorer health status and more use of hospital care (e.g. Palumbo *et al.* 2016; MacLeod *et al.* 2017). Among older adults, inadequate health literacy can be an independent risk factor for hospitalization (Baker *et al.* 2002). The results are contradictory, however, as other types of health literacy tests do not show any connections between lower scores and more use of hospital services (Jessup *et al.* 2017). Levy and Janke (2016) found more problems with accessing timely healthcare for adults aged 50 years or over with lower health literacy. Studies have, furthermore, pointed out that higher health literacy is also a substantial factor for adherence to treatments and medications, that is, to following recommendations or prescriptions (Lee *et al.* 2017; Soones *et al.* 2017).

Health literacy has been connected to perceptions of the quality of physician–patient communication as well. Patients with inadequate and marginal health literacy perceived the communication quality to be lower. They experienced problems with general clarity, explanations of their condition and processes of care, as well as empowerment, and decision-making. Patients with poorer health literacy are thought to be more likely to be confused or not enough informed about their own condition and the needed processes of care (Kripalani *et al.* 2010). MacLeod *et al.* (2017), in a similar fashion, found links between inadequate health literacy and dissatisfaction with the healthcare system as well as healthcare in general.

4.6. Health information literacy and health information behavior

Health information literacy is often studied in relation to information behavior. In an earlier publication containing results from the survey presented in this chapter, Eriksson-Backa *et al.* (2018) showed that higher levels of health information literacy were significantly related to stronger interest in and activity to seek health-related information. Enwald *et al.* (2018) found a link between more confidence in abilities to evaluate and understand health information and use of traditional and mobile information technology. Yang (2019) examined the role of health information literacy of older people in relation to health rumor sharing and found a strong relationship between higher health information literacy and lower intentions to share rumors. Eriksson-Backa and Nguyen (2020), furthermore, found strong relationships between everyday health information literacy and perceived intentions to be more active at seeking information in the case of a revealed higher risk for diabetes.

4.7. Measuring health information literacy

Compared to health literacy or eHealth literacy, health information literacy has been studied less extensively and generally with less standardized methods. In a qualitative study by Yates and colleagues (Yates *et al.* 2012), health information literacy was quite narrowly regarded as using information to learn about health. Eriksson-Backa (2010) surveyed everyday life health information literacy in a group of seniors in a quite general way by constructs measuring their needs for health-related information, their use of information sources, as well as their perceptions of information quality, understanding of information, and health knowledge and behavior.

Ivanitskaya and colleagues (Ivanitskaya *et al.* 2006), however, developed the Research Readiness Self-Assesment (RRSA) in order to measure the web searching skills of university students, as well as their ability to evaluate the obtained information about health, their knowledge about plagiarism, and their perceptions of their own seeking skills. Later, Niemelä *et al.* (2012) developed a ten-item everyday health information literacy screening tool (the EHIL scale). This scale has been the most extensively used scale so far, and has been applied on different populations (Nengomasha *et al.* 2015; Hirvonen *et al.* 2016; Enwald *et al.* 2018; Mayer 2018; Huvila *et al.* 2019; Eriksson-Backa and Nguyen 2020). Yang (2019) developed a new scale by combining the EHIL scale and a health literacy test called The Health Literacy Management Scale (HeLMS) (Jordan *et al.* 2013), whereas Wang *et al.* (2019) used the scale as a basis and added four items.

In our questionnaire, we assessed health information literacy using a test similar to the eHEALS (Norman and Skinner 2006). Eight statements covering needs, seeking, evaluation, understanding and use of health information were to be answered on a five-point Likert-type scale ranging from "totally agree" to "totally disagree". The statements were as follows:

– 1. "It is easy for me to decide in which kinds of situations I need health-related information";

– 2. "I know which sources to turn to in order to obtain health-related information";

– 3. "It is easy for me to find the health information I need from the information sources I use";

– 4. "I obtain too much health-related information";

– 5. "It is easy for me to determine whether health information is trustworthy or not";

– 6. "I learn many new things from the health-related information I obtain";

– 7. "I know how to use the health information I obtain to take care of my health"; and

– 8. "I often have difficulties to understand words or sentences used in health-related information".

In the analysis, "totally agree" was equivalent to five points and "totally disagree" to one point, except for the negatively worded statements 4 and 8 for which we reversed the points. The points were summed up and the minimum possible number of scores was hence 8 and the maximum possible scores 40 (see Niemelä *et al.* 2012; Hirvonen *et al.* 2016). We calculated health information literacy scores for 273 respondents altogether who had ticked all eight statements. The range of the total calculated scores was 11–40 with a mean score of 26.85 and a median of 27.

4.8. Communication between seniors and healthcare professionals

Level of health literacy has often been connected to satisfaction with health communication (e.g. Kripalani *et al.* 2010; MacLeod *et al.* 2017). To the best of our knowledge, no previous studies have linked health information literacy to patients' perceptions of communication in healthcare in a similar way, however. In this study, we looked at two aspects: how older adults perceived and experienced the communication in healthcare contexts, and if there were differences in the perceptions that could depend on either socio-demographic variables or health information literacy.

Previous studies have pointed to some problems that might occur in patient-provider communication. Lack of time during consultations is a factor that is often mentioned as a barrier in healthcare communication (Slort *et al.* 2011; Carollo 2015; Ting *et al.* 2016). Eriksson-Backa (2008) found that 7% of older adults often or always experienced that the healthcare provider had not given them as much information as they wanted due to lack of time, whereas 37% of them perceived that they had never experienced time constraints. Physicians' communication skills are also important in the communication situation, including both how the physicians behave towards the patients, and how easy they are to understand (Carollo 2015; Ting *et al.* 2016). Confusion due to contradictory advice has also shown to be a possible barrier. Hirvensalo *et al.* (2005) found that 34% of older patients had been given both recommendations for and warnings against physical exercise. Another study showed that 42% had at least occasionally obtained different advice from different physicians and 33% had obtained contradicting information concerning their medication. In the same study, it was found that only 39% had never felt that it was difficult to tell the physician if they did not understand the given advice (Eriksson-Backa 2008).

Sometimes information is withheld from the patients perhaps because physicians think that the patients do not even want to know. This might be the case especially for elderly or seriously ill patients (Giacalone *et al.* 2007; Slort *et al.* 2011). Posma *et al.* (2009) suggested that older patients should be encouraged to ask questions during appointments, so that the physician could get insights into their information needs. Peck (2011), however, found that practitioners do apply a more patient-centered style (as opposed to physician dominated) with older patients and that patient satisfaction is often higher among older adults over 65 years. Bo *et al.* (2014) drew the conclusion that although functional skills such as filling in forms, reading and understanding might deteriorate in older age, older people might still perceive the dialogue with healthcare providers to be of high quality. Older adults might, in fact, not even expect to be a participating partner with the health provider and hence not perceive problems in the communication with the provider.

We measured the perceptions and experiences of communication in healthcare contexts by eight statements that covered patients' own willingness to communicate with the practitioner, their opinion on the amount of given time and information, their perceptions of the clarity of obtained information, and their adherence to given advice. The statements were in part similar to the ones used in the study by Eriksson-Backa (2008). The eight statements were requested to be answered on a five-point Likert-type scale ranging from "totally agree" to "totally disagree". More positive perceptions or experiences would mean agreeing or totally agreeing to all but statement (*g*) (contradicting advice), which was negatively worded:

– (a) "It is easy for me to tell the physician that I have not understood what he/she said" (easy to tell).

– (b) "I ask my physician when I want more information e.g. about my health condition" (ask for information).

– (c) "Healthcare personnel have time to give me the information I want" (enough time).

– (d) "I understand from the prescription, the slip on the medication package or the patient information leaflet how often and in what way I should take my medication" (understand medication information).

– (e) "I follow the advice given by healthcare or pharmacy personnel on how to take my medication" (follow advice on medication).

– (f) "I follow the advice given by healthcare personnel on how to eat or live" (follow advice on behavior).

– (g) "I have been given contradicting advice from different physicians or other healthcare or pharmacy personnel" (contradicting advice).

– (h) "I have been content with the amount of information I was given during an examination or a treatment" (content with information).

Table 4.2 presents how the 273 respondents who were included in this study had ticked the statements about communication with health professionals. More than one-third (90/265) totally agreed with statement (a), that it is easy to tell the physician that they had not understood everything, and almost as many (30%, 80/265) agreed on this. Around 20% (53/265), however, disagreed or totally disagreed with the statement. More than one-half (143/269) totally agreed on asking the physician for more information if needed (statement (b)), whereas 26% (71/269) agreed. Concerning statement (c) about whether health professionals have enough time to inform the patients, the responses were dispersed; 25% (68/269) totally agreed, 36% (98/269) agreed, and 23% (62/269) were neutral. Only 15% (41/269) disagreed or totally disagreed. A mere majority of 73% (197/269) totally agreed, and an additional 20% (55/269) agreed to statement (d) (that they understand written information about their medication). Even higher percentages (nearly 78%, 208/268) totally agreed with statement (e) about adherence to advice about medications. In addition, 19% (51/268) agreed with this. When it comes to statement (f) and advice on how to eat or live, the respondents were less eager to follow advice, however. Fewer than 17.5% (47/268) totally agreed with the statement, whereas 40% (99/268) agreed. Nearly one-third (80/268) neither agreed nor disagreed. Concerning statement (g) about having obtained contradictory advice, only a small number of respondents could recall having experienced this: 4% (11/266) totally agreed, and 15% (40/266) agreed. As many as 40% (106/266) on the contrary totally disagreed with this statement, with an additional 27% (73/266) disagreeing. More than one-third (36%, 97/266) totally agreed with statement (h) about being content with the amount of information given during an examination or a treatment, and 42% (111/266) agreed. No respondents totally disagreed with this statement.

Statement	Totally agree	Agree	Neither agree nor disagree	Disagree	Totally disagree
(a) Easy to tell (n=265)	34.0%	30.2%	15.8%	10.9%	9.1%
(b) Ask for information (n=269)	53.2%	26.4%	11.9%	6.3%	2.2%
(c) Enough time (n=269)	25.3%	36.4%	23.0%	10.4%	4.8%
(d) Understand medication information (n=269)	73.2%	20.4%	5.2%	0.7%	0.4%
(e) Follow advice on medication (n=268)	77.6%	19.0%	2.6%	0.4%	0.4%
(f) Follow advice on behavior (n=268)	17.5%	39.9%	29.8%	8.6%	7.1%
(g) Contradicting advice (n=266)	4.1%	15.0%	13.5%	27.4%	39.8%
(h) Content with information (n=266)	36.5%	41.7%	15.4%	6.4%	0.0%

Table 4.2. *Responses to statements on communication in a healthcare context (adapted from Eriksson-Backa and Ek (2015, p. 81), reused with permission)*

4.9. Health information literacy and healthcare communication

By using standard multiple regression analysis, we tested the demographic variables gender, age group, and education level as well as calculated health information literacy scores against the eight statements about perceptions and experiences of communication within healthcare. We only included the 273 respondents who had ticked all eight statements measuring health information literacy, and set the significance levels at .05.

The regression analysis showed significant correlations between the six statements (*b*), (*c*), (*d*), (*e*), (*f*), and (*h*) and health information literacy scores on the $p<.001$ level. In addition, statement (*a*) was significantly correlated on the $p<.05$ level. Statement (*g*) did not show any significant correlations, however. Appendix 1, (Pearson r) shows these correlations. Health information literacy scores were, furthermore, significantly correlated with basic and high level education on the $p<.001$ level. The results of the

ANOVA show that the prediction model was statistically significant for all statements except (a) "easy to tell", where F(6, 256)=1.424, and p=.206, and (g) "contradicting advice", where F(6, 257)=.912, and p=.486. Appendix 1 shows the raw and standardized regression coefficients of the predictors and their correlations with the statements, as well as the squared semi-partial correlations.

The model was significant for statement (b) "ask for information", F(6, 260)=6.931, p<.001 and accounted for 12–14% of the variance of the statement (R^2=.138, Adjusted R^2=.118). Only higher scores of health information literacy was a statistically significant predictor for agreeing more with statement (b). In addition, the model was significant for statement (c) "enough time", where F(6, 260)=6.645 and p<.001. The model accounted for 11–13% of the variance of this statement (R^2=.133, Adjusted R^2=.113). To agree more with statement (c) was primarily predicted by higher health information literacy scores and to a lower extent by male gender. These two predictors were the only statistically significant ones. The model was also significant for statement (d) "understand medication information": F(6, 260)=4.616, p<.001. It accounted for 8–10% of the variance of statement (d) (R^2=.096, Adjusted R^2=.075). To agree more with statement (d) was predicted only by higher health information literacy scores.

For statement (e), that is, "follow advice on medication", the model was also significant with F(6, 259)=4.429, p<.001, and it accounted for 7–9% of the variance of the statement (R^2=.093, Adjusted R^2=.072). Agreeing more with statement (e) was on the first hand predicted by higher health information literacy scores, but to a lesser extent also by having a basic level education. These two predictors were the only statistically significant ones. In addition, statement (f) "follow advice on behavior" was also significant: F(6, 259)=4.014, p<.05. Here, the model accounted for 6–8% of the variance of statement (f) (R^2=.085, Adjusted R^2=.064) and agreeing more with statement (f) was predicted by only one statistically significant predictor, that is, higher health information literacy scores. Finally the last model, (h) "content with information", with F(6, 257)=6.834, p<.001, was also significant. In this case, the model accounted for 12–14% of the variance of the statement (R^2=.138, Adjusted R^2=.117). To agree more strongly with statement (h) was predicted by two statistically significant predictors, primarily by higher health information literacy scores and to a lesser extent by being a man.

The unique variance that was explained by each of the variables indexed by the squared semi-partial correlations was generally quite moderate as can be seen in Appendix 1. Health information literacy uniquely accounted for 11.2% for statement (b), 10.7% for statement (c), 6.8% for statement (d), 6.4% for statement (e), 6.2% for statement (f), and 10.8% for statement (h), whereas all other variables (statements) accounted for 3% or less.

4.10. Conclusions

Older adults largely prefer information from healthcare professionals or other medical sources, ahead of other types of sources, in both Finland and other countries. Because of this, it is crucial that the provided information is suitable for the receiver, and also that the receivers – the patients – are able to obtain, understand and use the information they obtain from these sources, that is, are health literate. An abundance of research shows that health literacy is associated with the use of health services and patient–practitioner communication. Health information literacy is a concept used more in everyday life than in mere healthcare contexts, and it is, in addition, much more focused on health information behavior and health information seeking skills. This concept is both newer and considerably less well studied than health literacy, possibly because it is largely limited to research connected to library and information science, whereas health literacy studies are conducted within the large areas of medicine, health sciences and health communication studies. Although a growing number of studies have more recently started to examine the health information literacy of older adults and its relation to acquisition of health-related information, no previous study has compared it to perceptions and experiences of healthcare communication.

This study found that although the respondents were mainly positive towards healthcare communication and eager to adhere to given advice, differences occurred that were related to health information literacy. The self-perceived abilities to seek and use health information were more strongly related to positive views of patient–practitioner communication than most demographic background variables were. In as many as six out of eight statements, the relationship between higher health information literacy scores and more positive experiences and perceptions of communication was significant. Thus, the results suggested that not only health literacy but also everyday life health information literacy seems to be an important factor for a positive healthcare communication experience and, eventually, a more healthy outcome.

4.11. Appendix 1: standard regression results

Statement (dependent variable)	Model	*b*	*SE-b*	Beta	Pearson *r*	*sr*^{2}
(a) Easy to tell						
	Constant	3.194	.471			
	HIL	.022	.016	.085	.107	.007
	Male	-.021	.161	-.088	-.026	.000
	Age 70–74	-.269	.183	-.097	-.080	.008
	Age 75–79	-.222	.213	-.069	-.041	.004
	Basic level	-.019	.184	-.007	-.077	.000
	High level	.289	.208	.094	.127	.007
	R^2=.032 Adjusted R^2=.010					
(b) Ask for information						
	Constant	2.300	.350			
	HIL*	.071	.012	.350	.364	.112
	Male	-.048	.121	-.023	-.057	.001
	Age 70–74	.118	.137	.053	.040	.002
	Age 75–79	.141	.161	.054	.029	.003
	Basic level	-.085	.139	-.039	-.116	.001
	High level	.019	.157	.008	.100	.000
	R^2=.138 Adjusted R^2=.010					
(c) Enough time						
	Constant	1.612	.374			
	HIL*	.074	.013	.343	.311	.107
	Male*	.392	.129	.177	.143	.031
	Age 70–74	-.138	.147	-.058	-.027	.003
	Age 75–79	.000	.172	.000	-.003	.000
	Basic level	-.010	.148	-.004	-.065	.000
	High level	-.161	.167	-.061	.018	.003
	R^2=.133 Adjusted R^2=.113					

Statement (dependent variable)	Model	*b*	*SE-b*	Beta	Pearson *r*	sr^2
(d) Understand medic information						
	Constant	3.831	.219			
	HIL*	.034	.008	.274	.284	.068
	Male	-.134	.076	-.105	-.131	.011
	Age 70–74	-.068	.086	-.050	-.034	.002
	Age 75–79	-.081	.101	-.050	-.027	.002
	Basic level	.035	.087	.026	-.063	.001
	High level	.050	.098	.033	.099	.001
	R^2=.096 Adjusted R^2=.075					
(e) Follow advice on medication						
	Constant	3.964	.197			
	HIL*	.029	.007	.265	.237	.064
	Male	-.109	.068	-.096	-.126	.009
	Age 70–74	-.105	.077	-.087	-.080	.007
	Age 75–79	.025	.090	.018	.061	.000
	Basic level*	.173	.078	.145	.067	.017
	High level	.012	.088	.009	.028	.000
	R^2=.093 Adjusted R^2=.072					
(f) Follow advice on behavior						
	Constant	2.116	.382			
	HIL*	.055	.013	.260	.276	.062
	Male	-.195	.132	-.089	-.112	.008
	Age 70–74	-.034	.149	-.015	-.013	.000
	Age 75–79	-.049	.175	-.018	-.012	.000
	Basic level	-.034	.151	-.015	-.089	.000
	High level	-.044	.170	.017	.090	.000
	R^2=.085 Adjusted R^2=.064					

Statement (dependent variable)	Model	*b*	*SE-b*	Beta	Pearson *r*	sr^2
(g) Contradicting advice						
	Constant	3.009	.440			
	HIL	-.025	.015	-.108	-.100	.011
	Male	-.065	.152	-.027	-.011	.001
	Age 70–74	-.049	.173	-.019	.002	.000
	Age 75–79	-.265	.202	-.087	-.079	.007
	Basic level	-.145	.175	-.057	-.024	.003
	High level	-.118	.196	-.041	-.039	.001
	R^2=.021. Adjusted R^2=-.002					
(h) Content with information						
	Constant	2.332	.297			
	HIL*	.059	.010	.345	.310	.108
	Male*	.268	.103	.153	.123	.023
	Age 70–74	.110	.117	.059	.055	.003
	Age 75–79	.242	.136	.110	.072	.011
	Basic level	.042	.118	.023	-.016	.000
	High level	-.144	.133	-.069	-.012	.004
	R^2=.138 Adjusted R^2=.117					

** $P<.05$. sr^2 is the squared semi-partial correlation. HIL=health information literacy (source: Eriksson-Backa and Ek (2015, pp. 83–85), reused with permission)*

4.12. References

American Library Association (2000). *Information Literacy Competency Standards for Higher Education*. American Library Association, Chicago. [Online]. Available at: http://www.ala.org/ala/mgrps/divs/acrl/standards/informationliteracycompetency.cfm [Accessed 17 June 2020].

Baker, D.W., Gazmararian, J.A., Sudano, J., Patterson, M. (2000). The association between age and health literacy among elderly persons. *J. Gerontol. B-Psychol.*, 55(6), S368–S374.

Baker, D.W., Gazmararian, J.A., Williams, M.V., Scott, T., Parker, R.M., Green, D., Ren, J., Peel, J. (2002). Functional health literacy and the risk of hospital admission among Medicare managed care enrolees. *Am. J. Public Health*, 92(8), 1278–1283.

Baron-Epel, O., Balin, L., Daniely, Z., Eidelman, S. (2007). Validation of a Hebrew health literacy test. *Patient Educ. Couns.*, 67(1–2), 235–239.

Berkowsky, R.W. and Czaja, S.J. (2018). Challenges associated with online health information seeking among older adults. In *Aging, Technology and Health*, Pak, R. and McLaughlin, A.C. (eds). Elsevier Academic Press, London.

Bo, A., Friis, K., Osborne, R.H., Maindal, H.T. (2014). National indicators of health literacy: Ability to understand health information and to engage actively with healthcare providers – A population-based survey among Danish adults. *BMC Public Health*, 14, 1095.

Carollo, S. (2015). Low health literacy in older women: The influence of patient–clinician relationships. *Geriatr. Nurs.*, 36(2 Suppl), S38–S42.

Digital and Population Data Services Agency (no date). Population information system [Online]. Available at: https://dvv.fi/en/population-information-system [Accessed 17 June 2020].

Enwald H., Hirvonen, N., Kangas, M., Keränen, N., Jämsä, T., Huvila, I., Korpelainen, R. (2018). Relationship between everyday health information literacy and attitudes towards mobile technology among older people. In *Information Literacy in the Workplace. ECIL 2017*, Kurbanoğlu, S., Boustany, J., Špiranec, S., Grassian, E., Mizrachi, D., Roy, L. (eds). Springer, Cham.

Eriksson-Backa, K. (2008). Access to health information: Perceptions of barriers among elderly in a language minority. *Inform. Res.*, 13(4), paper 368.

Eriksson-Backa, K. (2010). Elderly people, health information, and libraries: A small-scale study on seniors in a language minority. *Libri*, 60(2), 181–194.

Eriksson-Backa, K. (2011). Health information behaviour among retired Finns with either normal weight or obesity. In *ISHIMR 2011, Proceedings of the 15th International Symposium for Health Information Management Research, 8–9 September 2011, Zurich, Switzerland*, Bath, P.A., Mettler, T., Raptis, D.A., Sen, B.A. (eds). University of Zurich, Zurich, and University of St. Gallen, St. Gallen, and University of Sheffield, Sheffield.

Eriksson-Backa, K. and Ek, S. (2015). Health information literacy and communication in healthcare contexts: A study of older Finnish adults. In *ISHIMR 2015, The Seventeeth International Symposium for Health Information Management Research, York, UK, 24–26 June 2015, Health Informatics for Enhancing Health and Well-Being, Proceedings*, Bath, P.A., Spring, H., Sen, B. (eds). York St. John University, York, and University of Sheffield, Sheffield.

Eriksson-Backa, K. and Nguyen, H. (2020). Health information-seeking styles and health information literacy in relation to anticipated health-promoting behaviour – Results from an online diabetes risk test survey. *Int. J. Telemedicine Clin. Pract.*, 3(3), 192–208.

Eriksson-Backa, K., Enwald, H., Hirvonen, N., Huvila, I. (2018). Health information seeking, beliefs about abilities, and health behaviour among Finnish seniors. *J. Librarians. Inform. Sci.*, 50(3), 284–295.

Eronen, J. (2018). Terveyden lukutaito, terveys ja toimintakyky 75-vuotiailla jyväskyläläisillä [Health literacy, health and ability to act among 75-year-old inhabitants in Jyväskylä]. *Informaatiotutkimus*, 37(3), 25–28.

Giacalone, A., Blandino, M., Talamini, R., Bortolus, R., Spazzaban, S., Valentini M., Tirelli, U. (2007). What elderly cancer patients want to know? Differences among elderly and young patients. *Psychooncology*, 16(4), 365–370.

Hall, A., Bernhardt, J.M., Dodd, V. (2015). Older adults' use of online and offline sources of health information and constructs of reliance and self-efficacy for medical decision making. *J. Health Comm.*, 20(7), 751–758.

Hallows, K.M. (2013) Health information literacy and the elderly: Has the internet had an impact? *Ser. Libr.*, 65(1), 39–55.

Harris, R.M., Wathen, C.N., Fear, J.M. (2006). Searching for health information in rural Canada. Where do residents look for health information and what do they do when they find it? *Inform. Res.*, 12(1), paper 274.

Hirvensalo, M., Heikkinen, E., Lintunen, T., Rantanen, T. (2005). Recommendations for and warnings against physical activity given to older people by health care professionals. *Prev. Med.*, 41(1), 342–347.

Hirvonen, N., Ek, S., Niemelä, R., Pyky, R., Ahola, R., Korpelainen, R., Huotari, M.-L. (2016). Everyday health information literacy in relation to health behaviour and physical fitness: A population-based study among young men. *Libr. Inform. Sci. Res.*, 38(4), 308–318.

Huvila, I., Hirvonen, N., Enwald, H., Åhlfeldt, R.-M. (2019). Differences in health information literacy competencies among older adults, elderly and younger citizens. In *Information Literacy in Everyday Life. ECIL 2018*, Kurbanoğlu, S., Špiranec, S., Ünal, Y., Boustany, J., Huotari, M.L., Grassian, E., Mizrachi, D., Roy, L. (eds). Springer, Cham.

Ivanitskaya, L., O'Boyle, I., Casey, A.M. (2006). Health information literacy and competencies of information age students: Results from the interactive online research readiness self-assessment (RRSA). *J. Med. Internet Res.*, 8(2), e6.

Jessup, R.L., Osborne, R.H., Beauchamp, A., Bourne, A., Buchbinder, R. (2017). Health literacy of recently hospitalised patients: A cross-sectional survey using the Health Literacy Questionnaire (HLQ). *BMC Health Serv. Res.*, 17, 52.

Jordan J.E., Buchbinder, R., Briggs, A.M., Elsworth, G.R., Busija, L., Batterham, R., Osborne, R.H. (2013). The health literacy management scale (HeLMS): A measure of an individual's capacity to seek, understand and use health information within the healthcare setting. *Patient Educ. Couns.*, 91(2), 228–235.

Kloda, L.A. (2008). Health information literacy in Canadian medical curricula: An opportunity for librarians? *J. Hosp. Librariansh.*, 8(3), 314–322.

Kripalani, S., Jacobson, T.A., Ileko Mugalla, C., Cawthon, C., Niesner, K.J., Vaccarion, V. (2010). Health literacy and the quality of physician-patient communication during hospitalization. *J. Hosp. Med.*, 5(5), 269–275.

Lee, Y.-M., Yu, H.Y., You, M.-A., Son, Y.-J. (2017). Impact of health literacy on medication adherence in older people with chronic diseases. *Collegian*, 24(1), 11–18.

Levy, H. and Janke, A. (2016). Health literacy and access to care. *J. Health Comm.*, 21(Suppl), 43–50.

Li, Y. and Brilla, L. (2008). Improving health information literacy: An environmental health application. *Am. J. Health Ed.*, 39(4), 248–254.

MacLeod, S., Musich,, S., Gulyas, S., Cheng, Y., Tkatch, R., Cempellin, D., Bhattarai, G., Hawkins, K., Yeh, C. (2017). The impact of inadequate health literacy on patient satisfaction, healthcare utilization, and expenditures among older adults. *Geriatr. Nurs.*, 38(4), 334–341.

Martzoukou, K. and Sayyad Abdi, E. (2017). Towards an everyday life information literacy mind-set: A review of literature. *J. Doc.*, 73(4), 634–665.

Mayer, A.-K. (2018). Examining the factorial structure and validity of the everyday health information literacy screening tool. *Cogent Med.*, 5(1), 1–14.

Medlock, S., Eslami, S., Askari, M., Arts, D.L., Sent, D., de Rooij, S.E., Abu-Hanna, A. (2015). Health information–seeking behavior of seniors who use the internet: A survey. *J. Med. Internet Res.*, 17(1), e10.

Morey, O.T. (2007). Health information ties: Preliminary findings on the health information seeking behaviour of an African-American community. *Inform. Res.*, 12(2), paper 297.

Nengomasha, C., Abankwah, R., Uutoni, W., Pazvakawambwa, L. (2015). Health information literacy of University of Namibia students. *J. Stud. Humanit. Soc. Sci.*, 4(1–2), 179–192.

Neter, E. and Brainin, E. (2012). eHealth literacy: Extending the digital divide to the realm of health information. *J. Med. Internet Res.*, 14(1), e19.

Niemelä, R., Ek, S., Eriksson-Backa, K., Huotari, M.-L. (2012). A screening tool for assessing everyday health information literacy. *Libri*, 62, 125–134.

Norman, C.D. and Skinner, H.A. (2006). eHEALS: The ehealth literacy scale. *J. Med. Internet Res.*, 8(4), e27.

Osborne, R.H., Batterham, R.W., Elsworth, G.R., Hawkins, M., Buchbinder, R. (2013). The grounded psychometric development and initial validation of the Health Literacy Questionnaire (HLQ). *BMC Public Health*, 13, 658.

Palumbo, R., Annarumma, C., Adinolfi, P., Musella, M., Piscopo, G. (2016). The Italian health literacy project: Insights from the assessment of health literacy skills in Italy. *Health Policy*, 120(9), 1087–1094.

Parker, R.M., Baker, D.W., Williams. M.V., Nurss J.R. (1995). The test of functional health literacy in adults: A new instrument for measuring patients literacy skills. *J. Gen. Intern. Med.*, 10(10), 537–541,

Parker, D., Byng, R., Dickens, C., Kinsey, D., McCabe, R. (2020). Barriers and facilitators to GP–patient communication about emotional concerns in UK primary care: A systematic review. *Fam. Pract.*, cmaa002.

Peck, B.M. (2011). Age-related differences in doctor-patient interaction and patient satisfaction. *Curr. Gerontol. Geriat. Res.*, 2011, 1–10.

Ratzan, S.C. and Parker, R.M. (2006) Health literacy – Identification and response. *J. Health Comm.*, 11(8), 713–715.

Rudd, R.E. (2015). The evolving concept of health literacy: New directions for health literacy studies. *J. Commun. Healthc.*, 8(1), 7–9.

Sanchiz, M., Chin, J., Chevalier, A., Fu, W-T., Amadieu, F., He, J. (2017). Searching for information on the web: Impact of cognitive aging, prior domain knowledge and complexity of the search problems. *Inform. Process. Manag.*, 53(1), 281–294.

Schaeffer, D., Berens, E.-M., Vogt, D. (2017). Health literacy in the German population: Results of a representative survey. *Dtsch. Arztebl. Int.*, 114(4), 53–60.

Serper, M., Patzer, R.E., Curtis, L.M., Smith, S.G., O'Conor, R., Baker, D.W., Wolf, M.S. (2014). Health literacy, cognitive ability, and functional health status among older adults. *Health Serv. Res.*, 49(4), 1249–1267.

Shipman, J.P., Kurtz-Rossi, S., Funk, C.J. (2009). The health information literacy research project. *J. Med. Lib. Assoc.*, 97(4), 293–301.

Slort, W., Schweitzer, B.P., Blankenstein, A.H., Abarshi, E.A., Riphagen, I.I., Echteld, M.A., Aaronson, N.K., van der Horst, H., Deliens, L. (2011). Perceived barriers and facilitators for general practitioner–patient communication in palliative care: A systematic review. *Palliat. Med.*, 25(6), 613–629.

Soones, T.N., Lin, J.L., Wolf, M.S., O'Conor, R., Martynenko, M., Wisnivesky, J.P., Federman, A.D. (2017). Pathways linking health literacy, health beliefs, and cognition to medication adherence in older adults with asthma. *J. Allergy Clin. Immunol.*, 139(3), 804–809.

Sørensen, K., Van den Broucke, S., Pelikan, J.M. Fullam, J., Doyle, G., Slonska, Z., Kondilis, B., Stoffels, V., Osborne, R.H., Brand. H. (2013). Measuring health literacy in populations: Illuminating the design and development process of the European Health Literacy Survey Questionnaire (HLS-EU-Q). *BMC Public Health*, 13, 948.

Suri, V.R, Chang, Y.-K., Majid, S., Foo, S. (2014). Health information literacy of seniors: A review. In *Information literacy: Lifelong Learning and Digital Citizenship in the 21st Century, ECIL 2014*, Kurbanoğlu, S., Špiranec, S., Grassian, S., Mizrachi, D., Catts, R. (eds). Springer, Cham.

Tennant, B., Stellefson, M., Dodd, V., Chaney, B., Chaney, D., Paige, S., Alber, J. (2015). eHealth literacy and web 2.0 health information seeking behaviors among baby boomers and older adults. *J. Med. Internet Res.*, 17(3), e70.

Tiller, D., Herzog, B., Kluttig, A., Haerting, J. (2015). Health literacy in an urban elderly East-German population – Results from the population-based CARLA study. *BMC Public Health*, 15(1), 883.

Ting, X., Yong, B., Yin, L., Mi, T. (2016). Patient perception and the barriers to practicing patient-centered communication: A survey and in-depth interview of Chinese patients and physicians. *Patient Educ. Couns.*, 99(3), 364–369.

Trummer, U.F., Mueller, U.O., Nowak, P., Stidl, T., Pelikan, J.M. (2006). Does physician-patient communication that aims at empowering patients improve clinical outcome? A case study. *Patient Educ. Couns.*, 61, 299–306.

Wang, F., Luo, D., Sun, W., Yan, T., Hu, D., Pan, W., Luo, A. (2019). Health information literacy and barriers of online health information seeking among digital immigrants in rural China: A preliminary survey. *SAGE Open*, 9(2), 1–8.

Watkins, I. and Xie, B. (2014). eHealth literacy interventions for older adults: A systematic review of the literature. *J. Med. Internet Res.*, 16(11), e225.

van Weert, J.C., Jansen, J., Bensing, J.M. (2009). Older cancer patients information and support needs surrounding treatment: An evaluation through the eyes of patients, relatives and professionals. *BMC Nurs.*, 8, 1.

Wilkowska, W., Brauner, P., Ziefle, M. (2018). Rethinking technology development for older adults: A responsible research and innovation duty. In *Aging, Technology and Health*, Pak, R. and McLaughlin, A.C. (eds). Elsevier Academic Press, London.

Wilson, T.D. (2000). Human information behaviour. *Informing Sci.*, 3, 49–55.

Yang, M. (2019). Health information literacy of the older adults and their intention to share health rumors: An analysis from the perspective of socioemotional selectivity theory. In *Human Aspects of IT for the Aged Population. Social Media, Games and Assistive Environments. HCII 2019*, Zhou, J. and Salvendy, G. (eds). Springer, Cham.

Yates, C., Stoodley, I.D., Partridge, H.L., Bruce, C.S., Cooper, H., Day, G., Edwards, S.L. (2012). Exploring health information use by older Australians within everyday life. *Libr. Trends*, 60(3), 460–478.

5

Categorization of Audiences and Hierarchization of Actors in the Journalistic Narrative on HIV/AIDS Prevention. What Implications for Prevention Information?

5.1. Introduction

Since the end of the 2000s, prevention has diversified and new biomedical prevention tools[1] have been developed, including post-exposure treatment (PET)[2], treatment as prevention (TasP)[3] and pre-exposure prophylaxis (PrEP)[4]. These new strategies are contributing to a renewed conception of prevention. In particular, PrEP is helping to redefine the risk of HIV transmission and is challenging the norm of condom use [GIR 16, YOU 16]. It is recommended for audiences at high risk of acquiring HIV through sexual transmission, particularly men who have sex with men (MSM). In fact, the use of PrEP is part of moral debates, in a context marked by an increase in risky behaviors and a decrease in condom use, particularly among MSM [VEL 13]. Moreover, as these are more or less recent scientific

---

Chapter written by Cécile LORIATO.

1 The term "biomedical prevention" is used here to refer to all new biomedical prevention strategies, that is, those that use antiretroviral therapy and/or require medical intervention.

2 PET is also called an "emergency treatment": it is recommended within 48 hours of taking a risk.

3 TasP is when an HIV-positive person takes antiretroviral treatment. After a few months of treatment, the viral load becomes undetectable and HIV is no longer transmissible.

4 PrEP is a dual therapy used for HIV prevention by HIV-negative people.

discoveries intended for a "minority" public [LAP 19, p. 9], the information is controlled by the scientific and political actors and is poorly publicized. It should also be remembered that the media have played a primordial role in the emergence of AIDS as a public problem. They have participated in the construction of the "social AIDS phenomenon" – a notion used by the journalistic discourse – by making it exist in the consciousness of social actors, thus underlining their role in the "production of reality" [HER 88, p. 1110]. Through a process of categorization, they have contributed to the creation of prevention-focused audiences, legitimate in the public action against AIDS. The introduction of effective tritherapies at the end of the 1990s led to what has been described as a period of "normalization" in the fight against AIDS [SET 00]. During this period, the media lost interest in the subject, and so it was no longer a news item[5].

In this context, we seek to shed light on the way in which the journalistic narrative on biomedical HIV/AIDS prevention is constructed and what this implies for information about this subject. We consider here that the journalistic narrative is a "*socially legitimized and legitimizing 'scene of appearance'*" [QUÉ 92]. It is a dominant public arena [CEF 16, DOD 99] in which "*struggles for visibility*" [VOI 05a] of the AIDS problem, of the sick, of associations and of actions are engaged. The media, in this case the written press, through the narratives that they elaborate, shape visibility or invisibility according to "*procedures of selection and significant organization*" [VOI 05b, p.56]. They "*make the availability (of) statements in multiplied universes of reception possible and play the role not only of 'gatekeeper', but also of formatter and constructer of the discursive realities*" [VOI 05a, p.92]. Therefore, if the media are a "social mirror" that participate in "*what constitutes, deconstitutes, transforms the public space*" [CHA 97, p. 114], they are furthermore a space that participates in the symbolic constitution of causes and individual and collective subjects [VOI 05b]. The journalistic arena is thus "a discursive battlefield" within which a work of hierarchization is carried out, giving rise to effects of inclusion and exclusion of certain actors, actions, events or problems [DAL 16, p. 9]. The journalistic treatment of biomedical prevention, and PrEP in particular,

5 Here, an "event" is understood as a media construction: "Social events are not objects that can be found ready-made somewhere in reality and whose properties and avatars are made known to us afterwards by the media, with more or less fidelity. They exist only insofar as these media shape them". VERON E., *Construire l'événement*, Editions de Minuit, Paris, pp. 7–8, 1981, quoted in [HER 88, p. 1110].

shows the role that the media play in legitimizing causes or actors. In this chapter, we examine the way in which the journalistic narrative on HIV/AIDS prevention is constructed, focusing specifically on the categorization and prioritization of actors. How do these processes hinder the legitimization of PrEP? What are the implications for prevention information? We first review the methodology used, before characterizing the journalistic treatment of HIV/AIDS prevention. We focus on categorization as a process that can be stigmatizing, but which is also a public health necessity, and then we examine the evolution of the naming of users and populations at risk. Finally, we look at the hierarchization of the word of experts, politicians, users and associations within the journalistic narrative.

5.2. Methodology

The corpus analyzed in this chapter is composed of 98 articles that appeared in 11 publications in the French mainstream media between 2008 and 2017. The aim here is not to analyze the journalistic account of HIV/AIDS prevention as a whole, but to identify the elements that shed light on the issues involved in making biomedical prevention and its actors visible. Two indicators were defined. The first concerns the mobilization of categories defining exposure to HIV risk and designating a specific population. A second indicator is concerned with the mobilization of actors' voices within the journalistic narrative on biomedical prevention. In this case, we noted and counted the references to the actors' words in the form of reported speech, as well as in direct form. These two indicators were defined in a sequential perspective, according to three temporal sequences punctuated by scientific discoveries and political actions. Thus, the first sequence begins with the publication of the Hirshel report affirming the effectiveness of TasP. In France, this discovery led to the publication of the Lert-Pialoux report, which pointed to the need to expand the offer of prevention to include prevention treatment, and defined the orientations of the 2010–2014 National Plan to Fight HIV/AIDS and STIs. The second sequence starts in 2012, with the beginning of the ANRS Ipergay trial, which tests the efficacy of PrEP taken discontinuously or "on demand" – before or after risky sexual intercourse. The third sequence begins in 2014 with the results of the ANRS Ipergay trial showing the efficacy of on-demand PrEP, confirming the need to implement the recommendations on scaling up the prevention offer. This announcement led to a temporary recommendation for use (*recommendation temporaire d'utilisation*, RTU) in 2016 and then a marketing authorization (*authorisation de mise sur le marché*, AMM) the following year. This sequential approach sheds light on the evolution of the categorization of prevention audiences, as well as the place attributed to the different actors in the journalistic narrative.

5.3. Routinized journalistic treatment of HIV/AIDS prevention

The articles in our corpus concern prevention, which represents 6% of all journalistic coverage of HIV/AIDS between 2008 and 2017. The journalistic treatment of prevention is part of a routine that sheds light on the vestiges of a public health mission that the media were invested in the 1980s [CHA 94, p. 41]. Thus, the articles collected between 2008 and 2017 were published at specific times of the year, for example on the occasion of World AIDS Day, and are generally an opportunity for the media to give an epidemiological overview. There are a few articles offering analyses: 64 are factual articles, 27 are analyses and seven are interviews with experts. Most of them are the product of information manufactured by communication specialists within medical or public health institutions, for journalists who "favor readily available information from well-identified sources with institutional legitimacy" [AZZ 07, p. 124]. However, even in an article that repeats a press release verbatim, the political positioning of the newspaper title can be transparent. Observing the media coverage of AIDS at the beginning of the epidemic, M. Pollak notes that "*the words chosen to speak about AIDS and therefore automatically to speak about homosexuals show the extent to which medical information is dependent on the presumed moral conceptions of the readers*" [POL 88, p. 145]. Thus, despite the attenuation of the left–right divide in the media in the early 1980s [JUH 01, p. 187], the arrival of AIDS revived these oppositions. Although the media discourse has since been smoothed out by normalization, our corpus shows that right-wing press titles treat the subject in a more neutral way than left-wing ones: for example, there is only one analytical article for *Le Figaro* over the whole period analyzed, whereas *Libération* has 16. Variations in the way prevention is treated in the press can be explained by the social characteristics of journalists and, more specifically, by their interactions with the field of HIV/AIDS. If, at the very beginning of the appearance of AIDS, information on the subject was the prerogative of medical columns and specialized journalists, its rapid evolution into a "*social fact*" definitively changed the situation, also taking place on the economic, moral and cultural levels and leading to a multiplication of speakers [HER 88, p. 1122]. In our corpus, a few journalists are specialized in the field of HIV/AIDS, because of their militant or professional background. Only three of them write regularly. Moreover, 27 articles are anonymous. Beyond the normalized nature of the struggle, these considerations may help to explain the generally superficial treatment of prevention news.

5.4. Categorizing users: between stigmatization and the need for prevention

In their analysis of the media's construction of the "social AIDS phenomenon", C. Herzlich and J. Pierret observe that the use of terms such as "risk factor", "groups", "individuals", "populations" and "categories at risk" is becoming widespread to designate people who may be affected by the disease. In the articles analyzed, these terms, which come from the register of epidemiology, designate a homosexual "community" and "lifestyle", so that they appear to be "*intrinsically linked to the disease in a causal relationship*" [HER 88, p. 1121]. Their use contributes to the double stigmatization of the persons they designate by maintaining a link with the notions of "fault" or "danger" [THI 02, p. 50]. The categorization of the epidemic during the first years was a process of research and attribution of responsibility in a context where the origin of the disease, as well as the modes of transmission were still at the hypothesis stage. The information available at the time was subject to interpretations that tended to designate the "ideal" culprit-victim: the homosexual was punished for their unethical sexual orientation. However, the appearance of AIDS coincides with the recent recognition of the rights of homosexuals. By assuming their responsibility for the growth of the disease, the designation of homosexuals as a "risk category" contributed to delegitimizing them within society. O. Voirol notes that

> There is no guarantee that normative integration through narratives is necessarily successful, that is, that it gives a fair place to all members of the community and to all of the moral experiences that unfold within it. On the contrary, everything indicates that the semantic operations of media narratives contribute to 'distributing the cards' in a way that devalues and invisibilizes many [VOI 05b, p. 64].

Consequently, if it were lesser, the risk can take on a completely different reality for these individuals, who are then weakened by the accusation of responsibility for the growth of the disease, because of their identity rather than their practices. They can then put themselves in a position of "*counter-denunciation (denunciation of the denouncers), (of) denial or (of) paralysis*" [POL 87, pp. 77–78].

Subsequently, while still primarily referring to homosexuals – and sometimes to other stigmatized populations such as people who use drugs, prostitutes and people from sub-Saharan Africa – these categories are legitimized by scientific findings about modes of transmission, as well as by epidemiology, which indicates a greater number of cases among homosexuals. The fact that they are used by science – researchers, epidemiologists, doctors – contributes to the stigmatization of these people in the name of scientific evidence [COM 18, p. 6]. Medical risk is also perceived as a social risk and the term "risk group" becomes "*the mirror of the social conditions reserved for a particular category of people*" [POL 87, p. 78]. Tracing the genesis of the construction of the category of "risk group", Pollak concludes that:

> The nomenclature and epidemiological statistics not only provide the starting point for multiple medical reflections and research, but constitute an obligatory point of reference on the measures to be taken against the spread of the disease. In particular, the term 'risk group' or 'exposed group' invites those responsible for health policy to take specific preventive measures, taking into account the characteristics of the populations thus designated. While losing its initial moral connotations, this term, by force of circumstance, becomes the vector of prevention [POL 88, p. 130].

Thus, while categorizing the risk of HIV transmission and acquisition is disqualifying for part of the population, it also contributes to making groups and practices that require state action visible. The "at-risk groups" thus become the "public" for prevention. Always obeying a logic of hierarchization of all that makes up the social world, "*categorical struggles are at the heart of the modes of government to define who is entitled to what*" [COM 18, p. 7]. Categorizing according to epidemiological statistics allows for the implementation of specific measures for designated populations. Thus, the Lert-Pialoux Report warns of the need to direct prevention actions towards targets classified according to the prevalence of the groups that they belong to. Similarly, at the time of their introduction, TasP and PrEP were recommended for individuals taking risks, with high risk being an eligibility criterion for the ANRS Ipergay trial and then a condition for the prescription of PrEP during the RTU. It is therefore "*no longer only the risk itself, but its more or less voluntary or restrictive management, that becomes the major issue of controversy between scientists and the audience concerned*" [POL 87, p. 78]. Similarly, this audience becomes one of the main concerns of the media, which sometimes points to the lack of scientific research, political will and will on the part of users who do not adhere to preventive measures.

5.5. From a categorization of the public based on membership in a risk group to a focus on practices

In the corpus analyzed, we observe the recurrence of the use of homogeneous categories to define HIV risk. However, many of these articles are repetitions of press releases or results of epidemiological surveys that are not analyzed by the journalist. This leads to the interpretation that this use is more a matter of routinization of the discourse on HIV, rather than a desire to attribute responsibility to a group. If we look at the timing of the publication of articles using these categories, we see that they are mainly published on World AIDS Day, in an annual review of the results of the latest weekly epidemiological bulletins. The announcement of the start of the ANRS Ipergay trial initiated a transformation in the use of risk categories. In January 2012, in an article for Slate.fr, the journalist D. Lestrade wrote that PrEP was intended for "those who are most at risk", including "gay men who find it difficult to use condoms systematically". For the first time in the corpus, there is no question of a homogeneous category: the diversity of practices and individualities within the same epidemiological "group" is taken into account. Similarly, the journalist speaks of "high-risk individuals", without specifying the identity characteristics of these individuals: their common point here is risk-taking. In 2014, when the first results of the ANRS Ipergay trial were published, an AFP article reporting the words of Professor J-M. Molina, co-investigator of the trial, was picked up by many headlines. Although the scientist is only quoted in *Les Echos*[6], almost all of the articles use formulations close to the one used by the Professor, who at the time spoke of "homosexuals exposed to the risk of infection": "people with a high risk of contamination" to designate "men with significant sexual activity

6 Moutot Anaïs, "Sida : un traitement préventif 'à la demande' prouver son efficacité", *Les Echos*, 29 October 2014.

without condoms"[7], or "an HIV-negative person exposed to HIV through risky sex"[8], or later, in November 2015, "homosexuals with practices considered risky"[9]. Without denying the prevalence of HIV among men who have sex with men, the use of these expressions contributes to associating risk with sexual practices and not systematically with homosexuality.

Beginning in 2012, and more so in 2014, we begin to glimpse the role played by the journalistic narrative around PrEP in the perception of this new prevention method and its users. O. Voirol writes:

> On the one hand, the narrative can be an operator in maintaining places and positions in their instituted ordering: it then confirms social subjects in their functions, associated with 'types', and thus also reconducts despised positions in their postures of contempt. On the other hand, the narrative becomes liberating when it proceeds to the highly political gesture of questioning the places, practices, functions and reciprocal articulations between the different 'social types'. It is as if the narrative form, as social mediation, was permanently at the heart of these tensions within a community redefining itself normatively and practically in the conflict of recognition [VOI 05b, p. 69].

The evolution in the categorization of prevention users in the journalistic narrative, as well as their maintenance, reveals the tensions between these two functions of the narrative. On the one hand, the continued use of categories designating a homogeneous group as responsible for the increase in the number of new transmissions confirms the association of homosexuality and AIDS and contributes to defining PrEP as a tool reserved for homosexuals as the "type" at risk of acquiring HIV. By being presented as a means of prevention "not for everyone", the introduction of PrEP – and prevention treatment in general – disrupts the period of normalization of HIV, justifying D. Lestrade's concern: "*after years of decoupling LGBT issue from AIDS, gay men are clearly back to square one*"[10]. On the other hand, narratives defining PrEP users in terms of

7 "Prévention du Sida : un médicament à prendre avant un rapport sexuel prouve son efficacité", *Le Monde*, October 30, 2014.

8 Benkimoun, P., "Sida : la prévention en première ligne", *Le Monde*, November 6, 2014.

9 Proust, C., "Sida : feu vert à un traitement préventif", *Aujourd'hui en France*, November 24, 2015.

10 Lestrade, D., "PrEP : il existe un moyen de stopper l'épidémie de Sida, et ce n'est pas une bonne nouvelle pour les gays", *Slate.fr*, July 17, 2014.

their individual risk management, and not in terms of their identification with a risk category have a "liberating" function, particularly because they recognize previously disqualified users. However, this change in the categorization of users in no way puts an end to stigmatization. Indeed, personal narratives, like journalistic accounts, make and break social typifications. Similarly, this "liberating" function must be qualified because individual risk management implies that the responsibility for prevention lies with the individual: they are responsible for themselves and also for others. For example, responsibility for the increase in STIs is attributed to PrEP users because they do not use condoms, which contributes to dividing them, qualifying those who use condoms as the "good" users and disqualifying those who do not.

As early as 1982, M. Pollak showed that the term "homosexuality" is far from defining a homogeneous group. According to him, sexological analysis has made it possible to treat all sexual manifestations in the same way, bringing about a change in the scientific representation of sexuality by promoting the differentiation of representations in terms of sexual identities. Thus,

> These representations are at the origin of many 'groups' and 'movements' which claim a social space that is their own and allows, at the price of segregation, the blossoming of their sexuality. This logic of differentiation and segregation tends to weaken the 'strong' opposition between heterosexuals and homosexuals. It could produce a game of multiple and changing alliances in the struggle that concerns the classification of acceptable and unacceptable sexual practices [POL 82, p. 51].

Pollak's observation makes it possible to interpret another term used in the journalistic account, which covers the idea that sexual practices do not systematically define a sexual orientation. Indeed, the use of the expression "men who have sex with men", or "MSM", is part of the recognition of a category of people who do not all identify as homosexuals, but whose sexuality includes homosexual relations. This category became institutionalized with the fight against HIV/AIDS. It has been used since the mid-1990s in public health bulletins and in national HIV/AIDS prevention plans and programs. The use of this category of users, which helps to blur the boundaries between heterosexuality and homosexuality, is a major prevention issue. It recognizes, on the one hand, the high prevalence of HIV among men who have homosexual relations and, on the other hand, the existence of users to whom public health institutions do *not* address their prevention messages, whether in campaigns aimed at the general population or in community campaigns – the latter, moreover, being disseminated exclusively in places of homosexual sociability. However, "*this category is itself the bearer of renewed identity assignments, but its main limitation is that it neglects the social aspects of male-male sexuality, the representations associated with it and the strategies of affirmation or concealment that it gives rise to*" [CHA 13]. In the corpus analyzed, this term is not systematically used. It appears as much as the terms "gay men" or "homosexuals". Like the latter, it is associated with the categories of "at-risk group" or "at-risk population", thus creating a type of "at-risk" that is not based on a homogeneous community. Although it is difficult to distinguish when and why this category circulates in the media, the analysis of the corpus seems to illuminate a more regular use of the term "MSM" or "men who have sex with men" from the start of the ANRS Ipergay trial in 2012, when it previously seemed to be reserved for scientific publications.

5.6. Media coverage of biomedical prevention actors

5.6.1. *The word of the experts dominates*

Analysis of the journalistic account of biomedical prevention reveals the over-representation of certain actors, whom we refer to here as "experts". On average, they are present in more than 60% of the articles. Most of them are researchers in immunology, infectious diseases, virology or public health, specialists in HIV/AIDS, some of whom were at the heart of the discoveries about the epidemic in the 1980s. This is the case for Françoise Barré-Sinoussi, who was awarded the Nobel Prize in Medicine for the discovery of HIV, and Willy Rozenbaum, the first French doctor to have diagnosed a person with AIDS. Others have been at the heart of the fight for many years, such as sociologist, economist and epidemiologist France Lert, Professor Gilles Pialoux, who heads the infectious and tropical diseases department of a major Parisian hospital, and who co-investigated the ANRS Ipergay trial with Professor Jean-Michel Molina, an infectious diseases specialist. Others are researchers or doctors holding prestigious positions in international organizations, such as the World Health Organization (WHO) or UNAIDS, or in government agencies, such as ANRS. General practitioners involved in the fight against HIV/AIDS are also mentioned. All of the articles, with the exception of the tribunes and interviews with other actors, feature experts. According to Mathien, "*researchers are replacing, in one way or another, the information professionals that are journalists. They bring them – or are supposed to bring them – the guarantee of the quality of the content that they cannot,* a priori, *ensure themselves*"[11]. The discourse of the general information press thus devotes itself to the dominant and accredited social actors [AZZ 07, p. 115]. Thus, when they are not mobilized to present their research, the experts are solicited to comment on the news about HIV/AIDS. This is the case, for example, during international events or when a new study or epidemiological bulletin is published. The journalistic treatment of HIV/AIDS is therefore part of a medical, scientific and pragmatic approach, as in the early years of its appearance. In this regard, Pollak observes that

> the most eminent experts are often targeted and interviewed, not only in the narrow field of their competence, virology or immunology, but also on related social and political issues. (…) The importance of this alliance (between the media and the experts) is enough to distinguish AIDS from the historical epidemics to which it has been compared, with the scientific experts functioning as the exclusive source of technical and political legitimization, and the media playing the double role of instance of consecration and sounding board [POL 88, pp. 153–154].

In the corpus analyzed, we note a stronger mobilization of experts' words when news of research is announced: in 2008 during the International Conference on HIV/AIDS in Mexico City, in 2010 with the publication of the Lert-Pialoux report and the announcement that its recommendations would be included in the 2010–2014 National Plan, in 2012 with the start of the ANRS Ipergay trial, and in 2014 with the announcement of the effectiveness of PrEP on demand. While this observation may seem obvious, it highlights one of the construction mechanisms of the "*AIDS social phenomenon*" identified by C. Herzlich and J. Pierret: the "*event value*" attributed to research [HER 88, p. 1116]. If, at the beginning of the 1980s, the mediatization of scientific knowledge was a "*restitution of knowledge in the process of being constituted*" [HER 88, p. 1117], today it seems distinct in many respects. Indeed, there is not the same urgency to comment on current events concerning HIV/AIDS, but, above all, the institutionalization

11 MATHIEN M., Médias, santé, politique, Éditions L'Harmattan, Paris, 1999, p. 22, quoted in [AZZ 07, p. 115].

of scientific research on the subject has considerably modified its organization and its modes of financing. The media coverage of discoveries is thus less charged with economic stakes and visibility for the scientists. The announcements are made by means of communications sent to the media, which tends to standardize the information and, consequently, the points of view of the experts. Thus, there is a permanent consensus in the journalistic narrative. On the one hand, scientific information is rarely put into perspective with other research and even less with the stories of the people with whom and for whom it is conducted. On the other hand, the debates on the issues related to this research are not mediatized. According to N. Dodier, the framing of information by scientific institutions is characteristic of the modern epidemic regime. He writes: "*This struggle led by ethical or scientific institutions against the circulation of information not controlled by scientific journals, marks the very current vigor of the longer-term historical movement that aims to frame, through specialized arenas, the expression of opinions in the public sphere*" [DOD 99, p. 142]. This explains why the new biomedical prevention strategies have not been discussed in public arenas, but in specialized arenas.

5.6.2. *The voice of politicians is poorly represented*

In general, the voice of politicians is very weak in the articles of the corpus. More precisely, it is present in three articles out of the 17 that make up sequence 1, two articles out of the 10 in sequence 2 and 17 articles out of the 64 in sequence 3. These are reported speeches or press releases from these politicians. If we take a closer look at the moments when politicians' words are mobilized, the poverty of their interventions is even more obvious. For sequence 1, out of six articles concerning the Lert-Pialoux Report and the 2010–2014 National Plan, only one of them mentions Roselyne Bachelot, the Minister of Health at the time. Another article quotes fragments of speeches by several politicians at the Mexico City Conference, while a third is an article in *Libération* written by Jean-Louis Touraine, then a deputy from the Rhône region and vice-president of the AIDS study group at the National Assembly. For sequence 2, the words of the Minister of Health Marisol Touraine and the President of the Republic François Hollande are quoted on the occasion of World AIDS Day. For sequence 3, M. Touraine is quoted in the three articles announcing the RTU, as well as in one of the seven articles concerning the marketing of PrEP. Between these two announcements, the minister is cited in 10 of the 17 articles dealing with a controversy sparked by the posters displayed in November 2016, in the public space of *Santé Publique France*'s prevention campaign addressed to MSM and aimed at promoting the diversification of prevention means. It provoked negative reactions from political figures (mayors, deputies) who sometimes went so far as to censor the posters in their cities. Twelve of the 17 articles in section 3 report what politicians had to say about the campaign, without, however, focusing on its objectives: diversified prevention is never mentioned. The focus is on the reactions to the campaign, as well as the reactions of political figures who criticize it, with the media framing only touching on the political and societal issues raised by the visibility of homosexuality in the public space. Thus, some press titles only appear in the corpus at this time, such as the weekly magazines *Télérama* and *Marianne*, the latter having published five articles denouncing the homophobia of censorious mayors. This phenomenon increases the visibility of politicians' words without making them richer, since the same verbatim is quoted in different articles. This finding allows us to qualify the representation of politicians' speech in the corpus analyzed: it is visible in a succinct manner, not very varied and, in the case of reactions to the 2016 campaign, somewhat disconnected from HIV/AIDS prevention issues.

5.6.3. *AIDES, a key player in the association*

The associative actors are mobilized as prevention experts in the journalistic account. Their expertise is distinct from that of scientific experts, in that it is constituted by their proximity to the field and their exchanges with users. Often, they themselves have an intimate experience of HIV/AIDS: they are living with HIV, they are taking PrEP and/or have relatives living with HIV, or who have died of AIDS. While this expertise is based primarily on experience in the field of prevention, it also draws on the mediation of scientific work and the scientific work itself, as some association members also hold research positions in national agencies (notably the ANRS). In the journalistic account, however, they all appear as representatives of associations and only two articles mention the words of an actor designated as an AIDES activist – the latter, however, also being an employee of an international organization fighting HIV/AIDS, according to our sources. All of the others are employees in high positions in the associations' organization charts. Of the 27 association representatives appearing in the journalistic account, the association AIDES has 25. This domination of AIDES among the associative actors is questionable and the fact that it has been brought to light makes it possible to question its role in the legitimization of diversified prevention, and more particularly of PrEP. Indeed, historically, the association quickly became a partner of the state in the fight against AIDS. Moreover, the core group around which it was formed, bringing together academics, lawyers, journalists and doctors, has influenced the association's participation in the development of research programs within dedicated centers [PIN 02]. Thus, AIDES is a partner in national and international trials, such as the ANRS Ipergay trial. This integration into the world of research gives the association legitimacy to express itself on prevention and new scientific advances.

5.6.4. *The silence of the user of prevention*

The user of prevention is particularly under-represented in the articles in the corpus analyzed. In sequence 1, the voice of the user appears twice, on lemonde.fr, at the end of November 2009 and the beginning of December 2010, in the transcripts of question-and-answer sessions, the first being conducted with Gilles Pialoux as an expert on HIV/AIDS, and the second with Bruno Spire as president of the association AIDES. The headlines include a fragment of the interviewed expert's words, showing in both cases a concern about the state of the fight against HIV/AIDS. We have no socio-demographic information on the people asking the questions, who are Internet users and readers of lemonde.fr. Most of the questions are direct requests for the expert's opinion. They concern the possible end of AIDS, the relevance of the new National Plan to fight HIV/AIDS, the reasons for the increase in the number of new HIV-positive discoveries, advances in research and so on. In short, very general questions leading to consensual answers. In sequence 2, the user's voice is never mobilized. Although this is a sign of its rarity in the corpus as a whole, it is worth noting the few articles collected in this sequence. In sequence 3, the user is more solicited, although still very little in comparison to the experts. Thus, their word is mobilized in seven articles out of the 64 that make up sequence 3. In 2014, a report in *Le Monde* portrays a couple of serodivergent people. The journalist shows their story by putting the words of the young woman living with HIV in the foreground. In the last lines of the article, a scientific expert and an association leader are asked to confirm the effectiveness of TasP, thus legitimizing the story of the person living with HIV. In 2015 and 2016, two articles mobilized the voice of users: the testimonies of several PrEP users were used to illustrate the experts' statements and, in the 2016 article, the work of an association accompanier. In 2016, an article on the controversy surrounding SPF's prevention campaign gave a voice to some residents of a city where the posters were censored. The visibility of citizens' points of view allowed the journalist to illustrate the debate between politicians. This time, it was the word of citizens, users and potential users, that legitimized the decisions made. In 2017, a few months after Truvada was granted marketing authorization for PrEP, two articles mobilized the words of PrEP users in a succinct manner, always illustrative of an expert's word. A single article, in the form of a report, presents three PrEP users with different profiles. Their stories are reproduced in the form of quotes that the journalist does not comment on. The word of an expert is quoted; it is the general practitioner in charge of the medical follow-up of these three users. The article presents them as a companion in the appropriation of PrEP, an actor in the background, but obligatory to obtain a prescription and necessary for an adapted follow-up.

The words of the users in the articles analyzed are mostly confined to illustrating the words of experts. However, as numerous studies have shown, and in particular the work of J. Barbot on AIDS [BAR 02], the patient – and here also the user of prevention services – is an actor in the debate on health. They have also acquired a certain importance in scientific research, with greater involvement in trials. The ANRS Ipergay trial bears this principle in its acronym, which stands for "*Intervention Préventive de l'Exposition aux Risques avec et pour les Gays*". Nevertheless, the journalistic framing of PrEP does not show public participation in the research. The journalistic account favors the figure of the scientific expert or that of the association actor. The latter represents PLWHA and users of prevention. By giving priority to the voice of the association actor, the journalistic account confiscates the voice of the layman and confirms the existence of a hierarchy of voices. By making the users' experience visible through a scientific discourse, or the filter of a third party in the form of the association, it disqualifies lay expertise. As far as PrEP is concerned, the invisibilization of the voice of the user and potential user makes the use of this new means of prevention taboo. Indeed, the stories describing the PrEP users highlight their deviant characteristics: they insist on the rejection of condoms, on multi-partnership and on homosexuality. They thus contribute to the construction of a figure of the PrEP user that is limited by the scientific discourse, the accounts of associations and the angle chosen by the journalist to tackle the subject.

5.7. Conclusion

The aim of this analysis was to shed light on the way in which the journalistic narrative participates in the legitimization of biomedical prevention, and more specifically of PrEP, and what this implies for prevention information. It has been shown that new biomedical prevention strategies are poorly publicized and that the articles published during the period studied are mainly written by journalists, with little socialization in the field of HIV/AIDS. The analysis of the naming of prevention audiences revealed two main issues arising from their categorization. On the one hand, the designation of homosexuals as being responsible for the increase in the number of new HIV transmissions disturbs the so-called "normalization" of HIV/AIDS and defines PrEP as a tool reserved for them, since they are the ones most at risk. On the other hand, the definition of PrEP users according to their perception and individual management of risk, rather than their belonging to an epidemiological category "at risk", contributes to the inclusion in the journalistic narrative of individuals who do not define themselves as homosexuals, but men who have sex with men, actors who were not very visible until then. Furthermore, the interest in the representation of actors in the journalistic account sheds light on the construction of the legitimacy of PrEP. We have thus identified, in the words of A-M. Gingras, "primary definers". According to the political scientist, these are "*actors with an often prestigious status, who are able to impose their vision, if not on the entire press, then at least on a good majority of the major media*" [GIN 08, p. 5]. In the corpus analyzed here, scientific experts are among them: their point of view on diversified prevention, and then on PrEP, is predominant. A point of view on these prevention strategies thus prevails which, while it does not delegitimize them, does not tend to make them legitimate and therefore acceptable. The remarks concerning the uncertainties as to their effectiveness, the fears raised about abandoning condoms, the constant reminder that condoms are the best means of prevention, are made by HIV/AIDS experts, which tends to make them look very suspicious of these new strategies. At the same time, the users of these means of prevention are relegated to the margins of the journalistic narrative: they appear indirectly and in fixed representations, or even after the institutionalization of the issue. As a result, they are objects rather than subjects of prevention [AZZ 07, p. 125]. These power relationships that run through the public arena that is the journalistic narrative make the legitimization of biomedical prevention strategies and their users complex. However, the recommendations constructed within this reservoir of meaning become a subjective reality for social actors. The limited access to information and the homogeneity of the journalistic discourse on prevention can therefore be problematic for the appropriation of these new tools.

5.8. References

[AZZ 07] AZZEDINE L., BLANCHARD G., PONCIN C., "Le cancer dans la presse écrite d'information générale", *Questions de communication*, no. 11, pp. 111–127, 2007.

[BAR 02] BARBOT J., *Les malades en mouvement. La médecine et la science à l'épreuve du Sida*, Éditions Balland, Paris, 2002.

[CEF 16] CEFAÏ D., "Publics, problèmes publics, arènes publiques… Que nous apprend le pragmatisme ?", *Questions de communication*, no. 30, pp. 25–64, 2016.

[CHA 94] CHAMPAGNE P., MARCHETTI D., "L'information médicale sous contrainte. À propos du sang contaminé", *Actes de la recherche en sciences sociales*, nos 101–102, pp. 40–62, 1994.

[CHA 97] CHARAUDEAU P., *Le discours d'information médiatique : la construction du miroir social*, Nathan Institut National de l'Audiovisuel, Paris Bry-sur-Marne, 1997.

[CHA 13] CHARTRAIN C., DOURIS V., GIRARD G. *et al.*, "VIH/Sida : anciennes approches, nouveaux regards", *Genre, sexualité & société*, available at: https://journals.openedition.org/gss/2891, accessed 20 February 2018, spring 2013.

[COM 18] COMBY J.-B., PAGIS J., "Introduction. Politiques de catégorisation du monde social", *Politiques de communication*, vol. 1, no. 10, pp. 5–26, 2018.

[DAL 16] DALIBERT M., LAMY A., QUEMENER N., "Introduction", in DALIBERT, M., LAMY A., QUEMENER N. (eds), *Circulation et qualification des discours : conflictualités dans les espaces publics (1),* Études de communication, no. 47, pp. 7–20, 2016.

[DOD 99] DODIER N., "L'espace public de la recherche médicale. Autour de l'affaire de la ciclosporine", *Réseaux*, vol. 17, no. 95, pp. 107–154, 1999.

[GIN 08] GINGRAS A.-M., "La construction de la légitimité dans l'espace public", *Politique et Sociétés*, vol. 27, no. 2, pp. 3–9, 2008.

[GIR 13] GIRARD G., *Les homosexuels et le risque du Sida. Individu, communauté et prévention*, Presses Universitaires de Rennes, Rennes, 2013.

[GIR 16] GIRARD G., "Les réseaux mis en contexte : une perspective comparative de la prise en charge du VIH en France et au Québec", in BROSSARD B. and WHITE D. (eds), *La santé en réseaux : exploration des approches relationnelles dans la recherche sociale au Québec*, Presses de l'Université du Québec, Quebec, 2016.

[HER 88] HERZLICH C., PIERRET J., "Une maladie dans l'espace public. Le Sida dans 6 quotidiens français", *Annales, économie, sociétés, civilisations*, no. 5, pp. 1109–1134, 1988.

[JUH 01] JUHEM P., "Alternances politiques et transformations du champ de l'information en France après 1981", *Politix*, no. 56, pp. 185–208, 2001.

[LAP 19] LAPLANCHE-SERVIGNE S., SA VILAS BOAS M.-H., "Introduction. Les catégorisations des publics minorisés en questions", *Participations*, vol. 3, no. 25, pp. 5–31, 2019.

[PIN 02] PINELL P., *Une épidémie politique. La lutte contre le Sida en France, 1981–1996*, Presses universitaires de France, Paris, 2002.

[POL 82] POLLAK M., "L'homosexualité masculine, ou le bonheur dans le ghetto ?", *Communications*, no. 35, pp. 37–55, 1982.

[POL 87] POLLAK M., SCHILTZ M.-A., "Identité sociale et gestion d'un risque de santé. Les homosexuels face au Sida", *Actes de la recherche en sciences sociales*, no. 68, pp. 77–102, 1987.

[POL 88] POLLAK M., *Les homosexuels et le Sida. Sociologie d'une épidémie*, Métailié, Paris, 1988.

[QUE 92] QUERE L., "L'espace public : de la théorie politique à la métathéorie sociologique", *Quaderni*, no. 18, pp. 75–92, 1992.

[SET 00] SETBON M., "La normalisation paradoxale du Sida", in URFALINO P. (ed.), "Sida et action publique. Études réunies et présentées par Philippe Urfalino", *Revue française de sociologie*, vol. 1, no. 41, pp. 61–78, 2000.

[THI 02] THIAUDIERE C., *Sociologie du Sida*, La Découverte, Paris, 2002.

[VEL 13] VELTER A., SABONI L., BOUYSSOU A. *et al.*, "Comportements sexuels entre hommes à l'ère de la prévention combinée – résultats de l'Enquête presse gays et lesbiennes 2011", *BEH*, nos 39–40, pp. 510–516, 2013.

[VOI 05a] VOIROL O., "Les luttes pour la visibilité. Esquisse d'une problématique", *Réseaux*, vol. 1, nos 129–130, pp. 51–71, 2005.

[VOI 05b] VOIROL O., "Le travail normatif du narratif. Les enjeux de reconnaissance dans le récit médiatique", *Réseaux*, vol. 4, no. 132, pp. 51–71, 2005.

[YOU 16] YOUNG I., FLOWERS P., MCDAID L., "Can a pill prevent HIV? Negociating the biomedicalisation of HIV prevention", *Sociology of Health & Illness*, vol. 38, no. 3, pp. 411–425, 2016.

6

Historical Approach to Health Information in Anti-alcohol Prevention: The Case of the American National Prohibition Campaign (1910–1919)

6.1. Introduction

Listed as the third cause of avoidable death by *Santé Publique France*, alcohol is still considered to be "anchored in French culture"[1]. This ambivalence between a now widespread knowledge of the physiological consequences of alcohol and the difficulty for public action to legislate on private health behaviors, invites us to question the effectiveness of anti-alcohol prevention and in particular the impact of health information in modifying behavior [SCH 09].

We propose to put this question into perspective by analyzing the place of health information in an anti-alcohol prevention campaign that is both historically and geographically remote from the author: the campaign for national prohibition in the United States[2]. Prohibiting the sale, transport and manufacture of alcohol throughout the United States since 1919, this

Chapter written by Aude CHAUVIAT.

For a color version of all the figures in this chapter, see: www.iste.co.uk/paganelli/information.zip.

1 https://www.santepubliquefrance.fr/determinants-de-sante/alcool/notre-action.

2 The National Prohibition came into effect in 1919 by amendment to the constitution (Volstead Act) and ended in 1933 under the presidency of Franklin Roosevelt.

amendment to the American constitution was the result of a fight led over nearly a century by the temperance movement[3]. Nevertheless, the political and national scope of this law at the beginning of the 20th century and its final success are largely attributed to the effectiveness of the Anti-Saloon League's campaign[4], which flooded the country with prohibitionist literature and made Americans aware of the dangers of alcohol. At a time when beer and whisky were freely flowing in the country and when alcohol was still the object of popular beliefs – not all of which have disappeared – which attributed to it the power to give strength, stamina and virility, the question of the place of health information in prevention arose in all its ambivalence. At the time, it was a source of mistrust, unfamiliar and little understood by the population, yet it had unprecedented argumentative potential in a context of ignorance of the effects of alcohol consumption.

Prohibitionists quickly became aware of its potential, since health information played an increasingly important role in the public campaign of the Anti-Saloon League at the beginning of the 20th century. We show here that, far from limiting themselves to an argument of authority favoring the social acceptability of behavioral reforms, the stakes of health information are above all an issue of legitimization for actors from civil society to publicize their cause and to justify the imposition of norms on the general public. It is simultaneously a field of scientific knowledge, a field of informational and communicational skills and an instrument used to promote a normative and ideological discourse. Our research hypotheses defend, on the one hand, the idea that the prohibitionist campaign shows the shift from a moralizing approach of anti-alcohol prevention to health information. We postulate that this change is similar to a form of medicalization [AIA 98] of alcohol-related problems and constitutes a new form of "moral" normalization. On the other hand, we postulate that the use of health information within the prohibitionist campaign testifies to a renewed approach to prevention in a context marked by a redistribution of roles in the

3 Official term originally designating the members of the Temperance Societies which flourished in the United States in the 19th century and more widely individuals defending a religious way of life far from any form of excess, especially alcohol. In opposition to the prohibitionists, they advocate a moral and religious way of life rather than the political way of reform to put an end to the scourge of alcohol.

4 The Anti-Saloon League, founded in 1896 in Oberlin, Ohio by Howard Russell, is the organization credited with orchestrating the campaign (health and political) that led to the establishment of national prohibition by amendment to the US Constitution.

political and social domain and by ideological and scholarly quarrels between scientific temperance and hygienism.

Our analysis of the place of health information in the prohibitionist campaign focuses on the general public campaign of the Anti-Saloon League. To do so, we conducted a content analysis on a corpus of 201 pamphlets published and/or printed by the League between 1910 and 1917 before focusing on a sub-corpus of 82 pamphlets containing health information created by the Scientific Temperance Federation[5]. All of the communication materials used in this chapter come from the archives of the Anti-Saloon League Museum in Westerville, Ohio. In order to complete our analysis of the corpus with a historical perspective, we also analyzed primary sources such as the League's catalog of publications[6] and the correspondence of Cora Frances Stoddard[7], president of the Scientific Temperance Federation.

We will begin by characterizing the place of health information in our corpus by shedding contextual and historical light on the issues that underlie it. We will then approach health information in the corpus as popularized scientific information in order to analyze the mechanisms of transmission, proof and visibility [CHA 97, JAC 05]. Finally, we will evoke the instrumentalization of health information in the service of a profoundly ideological discourse by looking at the techniques of manipulation of health information and the implications of a necessarily subjective approach to this information.

6.2. Health information in the prohibitionist campaign

Health information currently occupies a crucial place in prevention, education and health promotion policies, particularly because it is part of a context of media over-information and multiplication of sources of health information. It is therefore necessary to orient and guide the public towards a

5 The Scientific Federation for Temperance was an association created in 1906 and specialized in scientific temperance instruction (physiological approach of anti-alcohol prevention). A dissident branch of the Women's Christian Temperance Union (WCTU), it worked closely with the Anti-Saloon League on the prohibitionist campaign.

6 Document from the archives of the Anti-Saloon League Museum, Westerville, OH.

7 The correspondence of Cora Frances Stoddard is held in the Special Collections of the New York Public Library.

reliable and legitimate discourse [KIV 14, CEC 08, FAV 05, ROM 15]. The historical context in which the League's campaign was situated was, on the contrary, marked by a rather rudimentary approach to health issues. Focused on tradition and interpersonal and generational transmission [RAU 13], it gave full scope to preconceived ideas and beliefs, particularly about alcohol. The latter also benefited from a strong moral stigma, confirmed by decades of activism on the part of temperance societies working for the moral salvation of the country through religion and the condemnation of all excesses. Thus, in spite of the decisive advances made by the scientific temperance movement[8] to educate and raise awareness among schoolchildren of the "physiological" dangers of alcohol from the second half of the 19th century onwards, the "general public" campaign was disseminated in a context where the moral approach remained dominant [BER 08, BER 07, DAR 06, GUS 81]. Health information was not very widely disseminated, and was even generally unknown to the general public, especially on this subject. Prohibitionists also came up against the consumption habits of the public for whom alcohol was an everyday substance – especially for the German or Irish immigrant populations – and finally against a counter-discourse widely relayed by anti-prohibitionists based on claims that alcohol nourishes, stimulates, strengthens, heals or refreshes. It is in this context that the question of the place of health information in anti-alcohol prevention, and more precisely in the prohibitionist campaign, arises.

6.2.1. *Quantifying health information*

6.2.1.1. *What is the place for health information?*

The corpus we are analyzing is made up of short communication media – one to two pages approximately – intended for the general public and in some cases reiterated in different printing formats: postcards for a large diffusion and poster format for display in a public space. Published between 1910 and 1917, these materials were mostly created by the Anti-Saloon League or the Scientific Federation for Temperance, then edited and printed in Westerville[9] before being distributed to order throughout the country.

8 Physiological approach to alcohol prevention.

9 In 1913, the Anti-Saloon League acquired an industrial printing plant in Westerville, which gave it the capacity to "flood" the country with prohibitionist literature.

We have concentrated our analysis of the corpus on five series of pamphlets, which are more substantial in terms of volume of items than some more confidential or highly specialized series. The dedicated term "series" designates a coherent and numbered set of pamphlets conceived according to a guiding principle (target, approach, aesthetic bias, etc.). We analyze here the "Cartoon" series, which proposes a very pictorial and rather moralizing approach to the question of alcohol; the "Safety First" series, which is centered on the prevention of the risks of accidents related to alcohol consumption in industrial and working-class contexts; the "Scientific Temperance Federation" or "STF" series, which proposes a scientific approach (physiological, sociological and statistical) to alcohol; the "Cherrington" series, composed of very short essays signed by Ernest H. Cherrington, superintendent of the League in Washington and head of publishing, and finally the "Alcohol" series, composed of anonymous essays on various topics related to alcohol.

There is a clear division between series that focus on health information and can therefore be described as "specialized" (Safety First and STF series) and those that do not address the issue from a health perspective (Cartoon and Cherrington series). The exception is the "Alcohol" series, which uses equal proportions of health information and other arguments. Nevertheless, we note that about half of the series uses health information, which tends to indicate that this "health" approach to the problem was in the process of acquiring real legitimacy. This first observation leads us to refocus our analysis on a sub-section of the corpus made up of the "Safety First" and "STF" series, that is, a total of 82 pamphlets which will henceforth constitute our main corpus. The "STF" series is, as its name indicates, a series created by the Scientific Federation for Temperance and printed by the League, as are some of the pamphlets making up the "Safety First" series. We will return later to the implications of this collaboration between the two organizations.

6.2.1.2. *Defining health information*

Our sub-corpus of pamphlets mobilizing health information is built on a discriminatory principle implying that it is possible to *define* health, a term with multiple and variable meanings depending on the time or context. Thus, where the WHO defines health in its entirety as "a state of complete physical, mental and social well-being, not merely the absence of disease or infirmity"[10], the question arises as to the relevance and anachronism of such a definition in a society that is still unfamiliar with the modern notion of well-being, whose medical and prophylactic practices are still in their infancy and whose sole objective is the absence of disease and infirmity.

The first definition of health information that we propose is built on the basis of a census of the themes addressed in the corpus and comes – somewhat schematically – to oppose the moral, political or economic argument. It is thus characterized by a pedagogical and popularized approach of human physiology as explained by medicine and more widely by science. Like recent work on the perception of health by the public [KIV 14], we note that it is a notion that is not very consensual, complex to grasp and very variable according to individuals, times and contexts. However, one of the common denominators of the different audiences is the direct or indirect involvement of the body [KIV 14]. The body thus constitutes the meeting point of the "health themes" encountered in the corpus. Nevertheless, this approach to the body is highly marked by resistance, endurance, performance and reliability. This rather "utilitarian" perspective also invites us to think of the body not in its individual and personal dimension but in its community dimension. The body is thus perceived by the American society of the time as one of the conditions of possibility for race welfare and for the prosperity of human societies, both through its reproductive functions and as a working tool. The summary in Table 6.1 sheds light on the thematic content of the pamphlets containing health information in the "Safety First" and "STF" series. They are categorized by types of effects on individual health, on the workplace and finally on family members, mainly children[11].

The category of individual health consequences tends to address the health effects of alcohol through the lens of performance. This approach invites qualification of the idea of a concern for the body rather than the individual. Individual health here refers to the maintenance of life and the absence of disease. These are very "medical" ways of approaching the relationship between alcohol and health, and they put us on the track of a

10 Preamble to the Constitution of the World Health Organization, as adopted by the International Health Conference, New York, June 19–22, 1946; signed July 22, 1946, by the representatives of 61 states. 1946 (Official Records of the World Health Organization, No. 2, p. 100) and entered into force on April 7, 1948.

11 The construction of the effect categories is based on a census of the themes addressed in the pamphlets mobilizing health information, completed by the observation of the phenomena of designation of victims.

pre-medicalization of health, understood as conferring a medical nature on realities that were not previously apprehended in these terms [AIA 98, p. 5]. This thematic census in fact reflects and is part of a very particular perception of health: that of scientific temperance.

Consequences of alcohol consumption	**Themes classified by order of frequency**
On **individual health**	– **Increase in mortality and decrease in life expectancy (32%)** – Immune deficiency with increased risk of infectious disease (tuberculosis, pneumonia, etc.) (16%) – Consequences on the nervous system and decrease of sensory acuity – Increased recovery and healing time after illness or injury
On **the job**	– **Increase in work-related accidents (30%)** – Decreased productivity and quality of worker performance (28%) – Extension of the duration of absence of workers due to illness or injury
On **family members**	– **Immune degeneration (60%)** – Learning disabilities and decreased academic performance

Table 6.1. *Summary table of health themes*

Scientific temperance has its origins in the tradition of Christian physiology, very popular in the North United States during the first half of the 19th century [ZIM 99]. The first health advocates in the United States put forward the idea that "a good knowledge of digestion and other bodily functions would help to keep away the vices of the flesh"[12]. Duly inspired by this precept, Mary Hunt, a leading figure in scientific temperance instruction in the United States, made a point of including in her definition of scientific temperance, "physiology and hygiene, with special reference to alcohol and other narcotics"[13]. Scientific temperance has the particularity of being a "scientific" movement carried exclusively by actors – and indeed actresses – coming from civil society and not from the academic world. Mary Hunt also defended the idea that it was up to non-specialists (lay people) to interpret and monitor the prescriptions of experts.

12 Sylvester Graham [ZIM 99, p. 120].

13 [ZIM 99, p. 120].

The three favorite topics of Scientific Temperance were: (1) the nature or composition of alcohol – leading to the conclusion that alcohol was a poison and not a food; (2) the dangers of alcohol on the organs: brain, bones, skin, liver, lungs, stomach, eyes; (3) the ability of alcohol to induce in children and adults alike constant craving and a diseased appetite – in other words, an addiction. The definition of health given in our corpus is thus modeled on that of scientific temperance and proposes health information centered on physiology. In this, it comes in opposition to a competing definition at the time, that of the hygienists, who bitterly defended the idea that it was necessary to inculcate healthy habits in children without encumbering them with physiological explanations justifying these good practices. They also did not want to reduce health to alcohol prevention, but to open it up to many other areas such as diet, exercise and hygiene. Health information as the Temperants conceived it addressed health through the prism of alcohol.

6.2.2. *The advent of health information*

Health information is to be considered in its contextual dimension as a reflection of a change of preventive paradigm. It became part of the "general public" campaign at the time when the prohibitionists decided to reorient their fight against alcohol around a single objective: to obtain national prohibition by amendment to the American Constitution. The presence of health information in the corpus also coincided with the institutionalization and structuring of a discipline and an international scientific community specialized around the issue of alcohol [BER 08, BER 07, DAR 06]. Health information finally found its place in prevention in an American society marked by a renewal of political ideology in the form of progressivism[14]. More concerned with sanitary and social problems – in particular questions of health, decency of living conditions and the fight against infectious diseases – American society at the beginning of the 20th century firmly believed in progress, progress that science and intellectuals in general were likely to bring.

14 Progressivism in the United States (1900–1915) was a political and ideological movement of the early 20th century that came in response to the observation that America was decrepit after decades of urbanization and industrialization and proposed to regenerate the country and revive American democracy through enthusiastic and relentless activism.

6.2.2.1. *The origins of health information*

Prevention is intimately linked to the notion of risk [PER 09] and the presence of health information in our corpus tells the story of the emergence of this notion. From the point of view of social history and literature, it is the passage from a "joyful" or "comic" alcohol to a "miserable" and "tragic" alcohol. It is with this radical change in perception that the need for prevention emerged [BAN 09]. Drunkenness becomes deviance and demands an investment of time, money, energy and consideration from the state.

The 19th century thus saw the flowering of campaigns that mobilized the principle of prevention as a tool for health progress. Rather coercive and more often concerning the way of life of communities than individual behaviors, these campaigns also tended to rely on shaky medical knowledge and far-fetched arguments.

The turn of the 20th century was a period when drunkenness was clearly perceived as incompatible with the family structure and the imperatives of the ongoing industrial revolution, and this inadequacy required legal action. The prohibitionist campaign was situated at the intersection between old preventive practices and modern prevention. A circumscribed but growing role was given to popularized health information as opposed to the moralizing approach. The traditional prohibitionist argument did indeed emphasize the social and economic consequences of alcohol and drew heavily on heart-rending stories of broken families and homes devastated by alcohol. However, as early as 1881, *The Alabama Baptist* newspaper mentioned the idea that the emotional register had little persuasive power [COK 07]. It occupied a place that was more or less identical to that of the informational register, and in particular to that of health information. This change is to be put in perspective with the emergence of scientific popularization [PAI 05] favored by the mass diffusion of printed materials. The emerging "scientific culture" intended to provide the public with information, as opposed to popular beliefs and rumors.

The advent of health information in the prohibitionist campaign can also be linked to the medical profession's quest for legitimacy, which began in the first half of the 19th century. American doctors began to adopt the principles of temperance around 1830 and became key players in an attempt to give professional status, credibility, respectability and legitimacy to their activity [MAR 08]. The advent of health information thus seems to be attributable to a favorable conjuncture: the emergence of the perception of alcohol as a social and health risk, the recognition of the limits of the emotional register in the field of anti-alcohol prevention, and finally the quest for legitimacy of the medical profession which found in the fight against alcohol an ideal vindication.

6.2.2.2. *Convergence of the international scientific community*

The summary table of health topics (see Table 6.1) discussed in the pamphlets shows the international dimension of the scientific perspective on alcohol. The themes of racial degeneration and the risk of immune suppression in the face of infectious disease were indeed commonplace in the scientific approach to alcohol in a variety of countries. The catalog of publications of the Anti-Saloon League mentions the existence of two booklets in the *International* Series entitled: "*The effects of alcohol on resistance to disease and offspring*" by Professor Taav Laitinen and "*The influence of alcohol upon the race*" by Professor Alfred Ploetz. The description of the second booklet refers to "a careful reflection on the old question of the effects of alcohol on the race, which emphasizes the hereditary deterioration effect of alcohol transmitted from generation to generation"[15]. Foreign scientists are also constantly used as references in our corpus of pamphlets and their experiences and works are often mobilized as scientific cautions.

This approach to alcohol and its consequences not on the individual but on the community and on the future of humanity in general was therefore part of a worldwide process of qualification and requalification of alcohol among the public [BER 08, BER 07, DAR 06]. The first phase is thus marked by an essentially moral qualification because the problem was taken in charge by temperance societies, both in the United States and in France. The end of the 19th century then saw the emergence in scientific discourse of a correlation between alcohol consumption and immune weakening in the face of infectious diseases (tuberculosis, syphilis and viral pneumonia). At the heart of this complex approach was the theory of the degeneration of the races of the Swedish doctor Magnus Huss[16], which would fuel until the middle of the Second World War the fear of transmitting alcohol-related

15 Pages 72 and 73 of the Anti-Saloon League publications catalog.

16 Magnus Huss, (1807–1890), Swedish physician who helped define *alcoholism* as a disease.

defects to future generations until only a degenerate population remained. For Bergeron, this was "the first attempt at a scientific qualification of alcoholism that links the pathology and its possible progression to the fate of the community" [BER 08, p. 343].

In the absence of any known medical treatment, only prophylactic measures were recommended, and yet this handling of the problem by scientific experts was a clear indication of the gradual decline of the influence of the temperance movements in dealing with the problem of alcohol, or of their transformation in the American context [BER 08].

6.2.3. *The acquisition of health information skills*

6.2.3.1. *From victims to experts*

The prohibitionist framing of the alcohol problem from the middle of the 19th century onwards treated alcohol above all as a threat to the home and to the education of children [COK 07, p. 207]. Thus, the initial "disorder" determining the framing of the alcohol problem [EME 77] – understood as the process of assigning the roles of complainant, victim and culprit – initially designated women and children as victims of alcohol and the saloon.

This status paradoxically allowed women to set themselves up as owners of the problem [GUS 81] and to make their voices heard in the public sphere, traditionally reserved for men. The year 1873 was thus marked by the popular episode of the "Women's Crusade" during which a total of about 50,000 women across the country engaged in militant forms of activism against saloons for nearly six months [KER 85, p. 44]. Their early involvement in this struggle would take on a decisive magnitude with the rise of the Women's Christian Temperance Union[17] (WCTU) created in 1874, directly following this seminal episode in the temperance struggle. Quickly converted to modern bureaucracy in full expansion, the WCTU structured its action in seven sectors supervised by a small number of women paid for these tasks, among which were social and preventive work, and also education in the form of scientific temperance instruction [KER 85, p. 47].

17 Women's Christian Temperance Union – a national women's association founded in 1873 and very active in the fight for temperance and prohibition.

6.2.3.2. *The Scientific Federation for Temperance*

After the death of Mary Hunt[18], an emblematic figure in the fight for scientific temperance education, uncertain relations with the Union of Christian Women for Temperance led the most fervent defenders of scientific temperance to create their own organization, the Scientific Federation for Temperance [ZIM 99]. Their project was to create an independent scientific institute whose vocation would be to collect the research work of the greatest experts on the subject and to confer legitimacy and authority on scientific temperance. They put forward the idea that "Physicians call everywhere for authoritative information and need such an organization" [ZIM 99, p. 123]. The Boston headquarters thus became a "research center" and the historical struggle to impose scientific temperance education in schools was replaced by a more diffuse objective, that of influencing American public opinion. Reorienting the Federation's activities towards the general public, Cora Frances Stoddard undertook to turn the Federation into an "office of expertise" [ZIM 99 p. 128] specialized in research, with little involvement in information dissemination activities, and enjoying the smooth and professional image of a new enterprise.

The fight for instruction by the Scientific Temperance movement and its slight reorientation under the guise of the Federation invites a political rereading of the notion of expertise. Far from being limited to a Manichean opposition between scientific expertise and citizen authority, the Temperance movement and its detractors alternately resorted to scientific discourse and to more popular beliefs. Health information as defended by scientific temperance was the business of those who could be described as intermediary experts, understood as (new) actors contributing to the publicization of a cause and benefiting from a legitimacy that justified their involvement in the production and dissemination of new social norms [BEA 85, DEO 14]. For the Temperants, this was an attempt to reconcile democracy and expertise in health [ZIM 99, p. 2]. The struggle for education in scientific temperance is not to be read as the acceptance or rejection of professional authority but rather as an enterprise carried out by citizens who – like the experts – used and selected health information to defend their own political agenda.

---

18 Mary Hunt (1830–1906): prominent temperance and prohibition activist, member of the WCTU and advocate for mandatory scientific temperance instruction in American public schools.

6.2.3.3. *Health information skills*

The management of health information in our corpus testifies to the appropriation and acquisition by the women of the Federation of real professional skills in health information and communication. The working relationships between the League and the Scientific Federation for Temperance evoke a specialization of the women of the Federation in the management and valorization of health information for prevention purposes and also the development of a form of expertise on the subject recognized by a variety of actors. To account for the acquisition of these skills by the women of the Federation, we analyzed the correspondence archives of Cora Frances Stoddard[19], president and chief editor of the Scientific Federation for Temperance. We also used the League's catalog of publications to account for the consideration given to the Federation's productions.

The communicative dimension of prevention and the skills required to handle health information enabled the Federation to establish itself as a "benchmark" organization in the health approach to alcohol control. It is this specialization that contributed to the legitimacy of the Federation's women in the political arena [DEO 14]. The example we are analyzing shows the emergence of a professionalization of information and communication practices within an organization originating from civil society and composed of women, whose place was traditionally confined to the private sphere. Thus, if the prohibitionist cause in general allowed them to speak out in the public sphere, it was their specialization in health information and more generally in the "scientific" approach of temperance that transformed them into resources in the prohibitionist fight, into professionals of health prevention. The health information sector is all the more interesting because it allows us to observe the emergence of specific professions based on the construction of specific skills [BOU 04]. These skills are built through action and require complementarity between pre-existing knowledge and knowledge acquired through practice.

These dispositions to explicitness and reflexivity that characterize social actors are systematized into competences [OLL 13], as testified by the correspondence of Cora Frances Stoddard, president of the Scientific

---

19 Cora Frances Stoddard, 1872–1936, prominent figure in the Christian Women's Temperance Union. Personal secretary to Mary Hunt, she became head of the Scientific Federation for Temperance on the latter's death in 1906 and remained at the head of this organization for 30 years.

Federation for Temperance. For Nathalie Berriau, "information professionals know not only how to select information but also how to analyze it [...]. They are able to give meaning, regularity, security and control in the processing of information for the benefit of the end user" [BER 16]. Table 6.2 identifies the types of tasks assumed by the Federation based on the description in the letters to Cora Frances Stoddard.

Information skills	Types of activities
Constitution of a thematic document archive	– Collection of documents in English and foreign languages – Document archiving – Request and collection of documents for the constitution of a fund (notably from public institutions) – Collection of documents sent by supporters for collection/consultation/inspiration
Quantitative and qualitative data collection activities	– Collection of public and private statistical data (public institutions, insurance companies) – Setting up survey protocols for statistical data collection (conducting interviews by correspondence and distribution of questionnaires)
Scientific popularization and data formatting activities	– Publication activities (books, media, journals and slides) – Editing and layout activities – Adaptation of content to the type of audience (age, profession, etc.) – Statistical data analysis and formatting
Valuation and dissemination of health information	– Participation in conferences as speaker(s) – Written contributions to publications (newspapers/encyclopedias) upon request of authors/publishers – Organization of fixed and mobile exhibitions throughout the country – Creation of media (graphics and posters) for public display and colored slides for mobile projection (on order) – Provision of projection equipment – Coordination of prevention activities with the Anti-Saloon League – Canvassing activities and service offers
Support for the search for information	– Management and centralization of document requests – Assistance in finding information – Sending of document resources – Document reference requests for third parties – Sending raw statistical data – Sending of formatted materials
Document expertise	– Publication of reference books/official works – Recognition of the informational/scientific/aesthetic quality of publications (by peers, by readers and by public institutions) – Securing the intellectual property of the Federation's work – Awards for Federation publications

Table 6.2. *Identification of the Federation's health information competencies*

The analysis of Cora Frances Stoddard's correspondence and of the League's catalog of publications indicates that the presence of health information corresponds to the development and construction of real informational skills by the members of the Federation. This specialization was based on the acquisition by the women of the Federation of the status of experts in health information and more generally in scientific information, making them references in the field of popularization. However, information – in this case health information – remains a subjective construction influenced in particular by cognitive, affective and situational factors that influence the organization of knowledge [CHO 00]. The apparent objectivity conferred by health information on the Federation's publications must in no way mask the intervention of the issuer in the collection, organization and formatting of the data. It would be imprudent to separate health information from the white Protestant middle-class milieu to which the activists of the Christian Women's Temperance Union belonged, and from the very moral and religious dimension that characterized it.

6.3. Communicating health information

The second part explores the informative dimension of the corpus through the prism of the ways in which health information is processed and transmitted. This leads us to evoke at the same time the argumentative dimension of the pamphlets. Indeed, while the role of informative discourse is to provide the public with knowledge, the function of argumentative discourse is to alter beliefs and mental representations [COM 88]. There is thus behind the objective of active prevention of the dangers of alcohol, a political objective of convincing the whole population to support the ratification of the amendment for national prohibition.

6.3.1. *Comparison of the series*

The two series we wish to analyze in order to report on the use of health information in our corpus are the "Safety First" series (36 pamphlets) and the "Scientific Temperance Federation" or "STF" series (58 pamphlets).

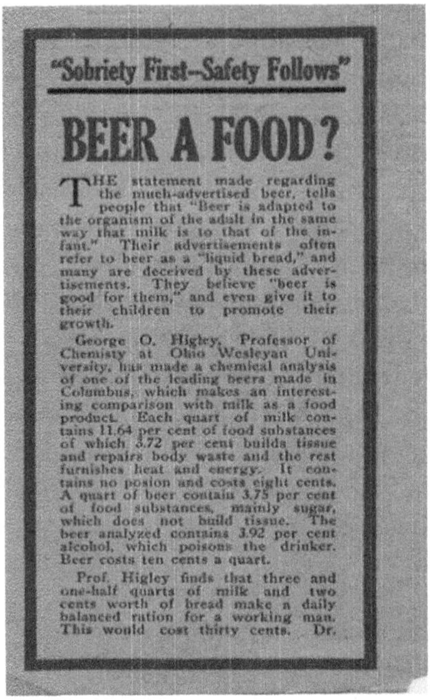

"Sobriety First—Safety Follows"

BEER A FOOD?

THE statement made regarding the much-advertised beer, tells people that "Beer is adapted to the organism of the adult in the same way that milk is to that of the infant." Their advertisements often refer to beer as a "liquid bread," and many are deceived by these advertisements. They believe "beer is good for them," and even give it to their children to promote their growth.

George O. Higley, Professor of Chemisty at Ohio Wesleyan University, has made a chemical analysis of one of the leading beers made in Columbus, which makes an interesting comparison with milk as a food product. Each quart of milk contains 11.64 per cent of food substances of which 3.72 per cent builds tissue and repairs body waste and the rest furnishes heat and energy. It contains no posion and costs eight cents. A quart of beer contain 3.75 per cent of food substances, mainly sugar, which does not build tissue. The beer analyzed contains 3.92 per cent alcohol, which poisons the drinker. Beer costs ten cents a quart.

Prof. Higley finds that three and one-half quarts of milk and two cents worth of bread make a daily balanced ration for a working man. This would cost thirty cents. Dr.

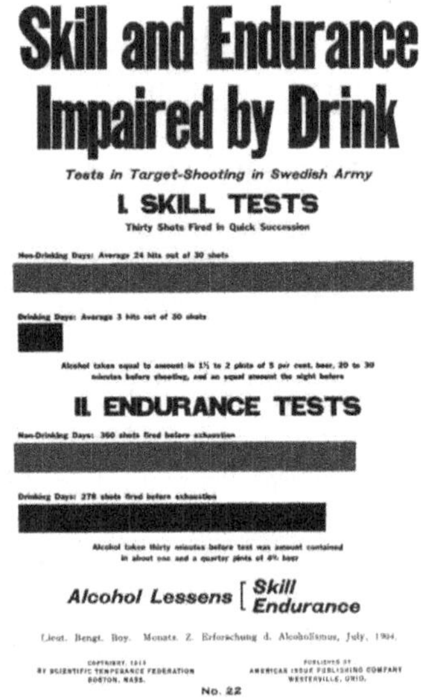

Figures 6.1 and 6.2. *Pamphlet "Beer a food?" (Safety First) and Pamphlet 22 (STF)*

With 84% and 92% of the pamphlets containing health information, respectively, these two series specializing in the physiological approach to the problem of alcohol have differences and similarities that need to be highlighted before taking a closer look at the way in which they mobilize health information. Starting with the differences, we can see that bias in the form of target audience segmentation is very different in each series. For example, the "Safety First" series claimed to be specifically designed and used in the context of "various industrial issues" and was distributed to workers in their pay envelope. It was also available in 12 foreign languages for immigrant workers.[20] This series, issued by the League's Department of Safety and Efficiency and the Scientific Federation for Temperance (working together on 11 pamphlets), was a specific response to the increased perception of the risks of intoxication in a context of rapid mechanization

20 Pamphlets available in French, German, Greek, Hungarian, Italian, Lithuanian, Norwegian, Polish, Romanian, Russian, Spanish and Swedish.

and industrialization of the country and the widespread use of dangerous, complex and sometimes deadly machinery. In contrast, the "STF" series issued by the Scientific Federation for Temperance adopted a strategy of "general public" diffusion without particular segmentation and with the objective of global dissemination of the information and saturation of the public space (see formats dedicated to public display).

This difference in strategy in terms of audience segmentation and dissemination had directly observable consequences on the media. The "Safety First" series opted for two-page pamphlets, relying mainly on written and detailed information with an average of about 320 words per pamphlet. The "STF" series, on the contrary, placed great emphasis on the visual transmission of information. All of the pamphlets offered graphic representation (graphs, diagrams, illustrations and pictorial graphics) with textual content often limited to a title and a quick commentary on the poster's visual. Designed to be attractive and quickly assimilated by a general public with low cognitive availability – although this is debatable – the strategies for transmitting health information were therefore based on very different codes for each series.

The shift to a preventive discourse based on health information marked an evolution towards a *medical* approach to health, also made possible by the acquisition of legitimacy of scientific discourse among the general public, notably through popularization [PAI 05]. The preventive strategy adopted in our corpus thus consists of educating and sensitizing the public to the dangers of alcohol by providing them with basic sanitary and physiological information in order to modify their state of knowledge and possibly their behavior. In this configuration, informing or transmitting health information implies making knowledge accessible in several ways so that it can be received, defined, understood and interpreted [CHA 97 p. 41].

According to Daniel Jacobi [JAC 05], popularized discourse is a reformulation of scientific discourse and in particular of the specialized vocabulary that characterizes this expert discourse. Our corpus thus shows a variety of definitions based mainly on paraphrase, reformulation or exemplification. The information provided is elementary and easily accessible, testifying to a state of knowledge of the general public that is not yet sufficiently advanced to tolerate nuances and complexities. The specialized vocabulary is also reformulated by means of synonyms and hyperonyms and generally relies on analogy and comparison to facilitate the assimilation of definitional content. The corpus nevertheless shows a certain poverty of vocabulary and favors the repetition of known terms from one sentence to the next in order to guarantee as much coherence as possible without concern for style.

Excerpt 1: "Some disease germs, like those of pneumonia and diphtheria, throw off poisons from their bodies. These poisons enter the blood and are carried to the cells of the body, which they poison until we die. When these poisons are in the blood of a healthy person, a substance will come into the blood which will help destroy the poisons and we get well more quickly". "How we catch disease" (D27 – Safety First).

The lexical redundancy of our corpus invites us to take an interest in the informational structure of the pamphlets and in particular in the passages whose function is to provide health information or to explain it. It is indeed possible to analyze informative texts according to a cognitive approach since preventive discourses or popularization discourses aim at informing the reader [JAC 05]. Combettes and Tomassone [COM 88] thus insist on the importance of analyzing the "shared knowledge" (what is known and what is not) as well as the density of information of a text and the balance between the quantity of known information and the quantity to be transmitted. The pamphlets in our corpus are, for the most part, short and not very dense texts. Echoing the quarrel between hygienists and Temperants, the objective for the latter is not simply to inculcate good behavior without explaining the cause, but to transmit just enough information to convince and not just impose. Combettes and Tomassone [COM 88] evoke on this subject the "rule of progression" which requires that an informative text gives the reader constantly renewed information and that he or she be able to identify this progression themself. Generally speaking, the general informational structure of pamphlets follows a fairly strict order. The first step is the description of an imperfect or erroneous state of knowledge. The second step consists of referring to the correct information and proposing its verification by means of tests, experiments or by relying on pre-existing scientific protocols. The last step confirms the hypothesis and confirms the new way of understanding and explaining reality. The repetition of the lexicon, the connectors and the temporal markers are used to materialize this progression.

6.3.2. *Health information and evidence regime*

According to Charaudeau [CHA 97], the informative text is based on the postulate of the public's ignorance, especially in the field of health information. Nevertheless, this ignorance may be of a variable nature and therefore require a differentiated approach. At the heart of the quarrel between hygienists and advocates of temperance was a profound disagreement between the desire to inculcate good behavior and the need to explain it, between the justifications for a healthy lifestyle and the simple rules that governed it [ZIM 99, p. 122]. The Temperance position – because it claimed the need for knowledge (as opposed to belief) – involved consideration of truth and evidence.

6.3.2.1. *The rhetoric of error*

In a context marked by preconceived ideas about alcohol and its effects and by high consumption rates among the population, the question of ignorance is relevant. This is indeed a crucial issue for analyzing the place of health information in our corpus and for understanding the difference between "*knowledge-based expertise*" and "*knowledge of belief*" [CHA 97 p. 44]. According to Charaudeau, "*knowledge-based expertise*" derives from a rational approach to the world and is based on scientific data and experiments, calculations, reasoning and explanations. It is meant to be detached from any social or cultural influence. Conversely, "*knowledge of belief*" is the result of a subjective "commentary" on the world, based on judgment and evaluation.

Pamphlet	
DON'T STAY SICK	Some people make it easier for themselves to get sick by drinking alcoholic liquors **although they think** this makes them stronger.
HOW TO LIVE LONG	Doctors **used to think** that drinks like beer and brandy made well people stronger and sick people well.
DANGERS FROM INCREASED HEAT	**The idea that** alcoholic liquors help one stand heat **is simply due to** the fact that alcohol by dulling the feelings **deceives** one as to its real effects.

Table 6.3. *The rhetoric of error*

This distinction sheds a great deal of light on the use of health information in the sense that the challenge is to get the public to adhere to a scientific truth, to a knowledge of knowledge, where knowledge of belief has always been considered as truth. A common rhetorical strategy in the "Safety First" series is to point out false beliefs and to designate them as erroneous before re-establishing a scientific truth. Table 6.3 offers some examples.

The passages in bold have the function of invalidating the readership's beliefs and designating them as false beliefs. Lexically, this strategy is reflected in the use of the verbs "to think" and "to believe" and in a strong use of the vocabulary of error. This rhetorical strategy makes it possible to move from the realm of belief to the realm of knowledge and tends to reinforce the idea for the reader that this is an informative text. Often located at the beginning of pamphlets to attract attention, this technique of destabilizing the reader encourages him or her to re-evaluate his or her beliefs and to take an active part in the process of assimilating the health information.

The "STF" series adopted a somewhat different strategy in that the error or mistaken belief was not explicitly named most of the time. Nevertheless, it is noticeable that the titles of the pamphlets – which often took up more than a third of the poster – addressed popular prejudices about alcohol, such as the idea that alcoholic beverages gave the consumer strength, energy and dynamism and improved his or her performance at work, and also ideas that alcohol was harmless to children born or unborn. Instead of deconstructing erroneous beliefs, this series opted for the imposition of proven knowledge, in particular through the use of very little moderation, similar to the results of scientific experiments and accentuated by the presence of graphs – and therefore difficult to contest.

6.3.2.2. *The rhetoric of expertise*

At the heart of Mary Hunt's fight for scientific temperance education was the idea that it was the duty of citizens to appropriate scientific information and that expertise was not the prerogative of scientists. Such appropriation of health information raises two questions: what is the value of (the) truth disseminated by non-specialists and does it claim the same type of rigor and objectivity as scientific truth [ZIM 99]? This questioning is at the heart of the opposition between "*truth effect*" and "*truth value*" [CHA 97, p. 44].

Charaudeau defines the "*value of truth*" as "an explanatory construction that is elaborated with the help of a scientific instrumentation, supposed to be external to man (even if he is the one who has constructed it), objectifying and objectified, which can be defined as a set of techniques of knowing how to speak, of knowing how to comment on the world". Conversely, the "*truth effect*" stems from a subjective approach to the world that determines what we believe to be true and that relies on social norms to define what can be "considered" as true and therefore shared with the rest of society. It is rather a matter of conviction and hides under the guise of knowledge what is above all opinion.

Charaudeau puts forward the idea that informative texts use the "*truth effect*" in different ways depending on the reasons for which the information is transmitted (motives), the social and psychological characteristics of the person giving the information (identity) and finally the means used to prove the veracity of the information (proof). As far as our corpus is concerned, the two main informants were the Anti-Saloon League and the Scientific Federation for Temperance, which can be equated to specialized organizations according to Charaudeau's typology [CHA 97] and whose stated ambition was to set themselves up as reference organizations for the prohibitionist cause. The identity of the speaker was nevertheless very little claimed in the pamphlets which were often anonymous and whose editors were designated very discreetly. On the contrary, the credibility of the information relied here on the prominence of the referenced authors, who were most often eminent scientists – doctors and professors with an international reputation. Seventy-nine percent of the pamphlets in our corpus made use of one or more references to the work of doctors, scientists (referred to as professors) or reports published by international organizations specializing in health, insurance, safety or alcoholism. This percentage tends to specify the scientific or medical anchoring of our corpus and to confirm the idea of an ongoing medicalization of the alcohol issue [AIA 98, p. 5]. Through this system of references, the sender of the message becomes the transmitter of health information whose validity is not to be demonstrated because it comes from a reliable source, legitimized by his status and his profession. This positioning also masks the intervention of the sender of the message in the information transmission process. This phenomenon is closely linked to the notion of expertise as described by Léglise and Garric [LÉG 97, p. 3]. The latter define expertise as discursively marked by a certain number of rhetorical procedures that our corpus allows us to observe, such as a relative absence of enunciation markers, the frequent use of passive or impersonal turns of phrase, and finally the presence of verbs used for description.

Since popularized scientific discourse is closely linked to the notion of truth [CHA 97], it is also linked to the notion of proof[21], which attests to the veracity of the informative discourse. These validation criteria aim to guarantee the plausibility or the credibility of the information by providing explanations on the origin of the knowledge. They also prove the authenticity of the information through graphs, examples or images, bringing a technical and scientific dimension to the debate.

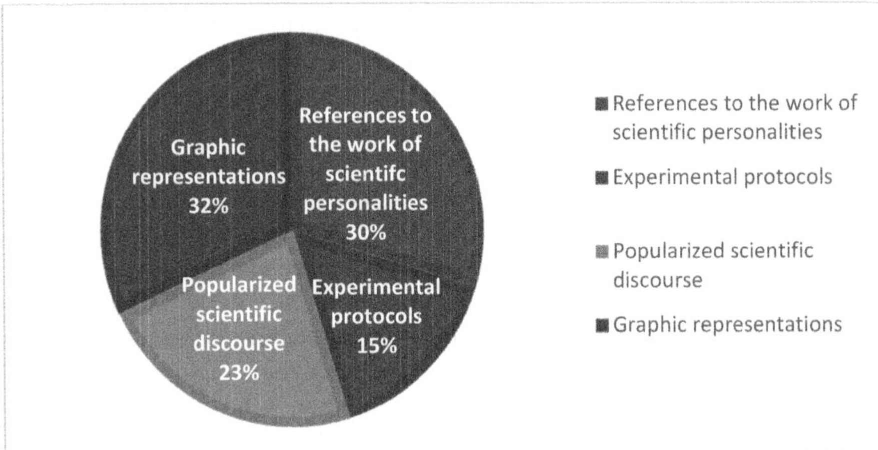

Figure 6.3. *Modalities of evidence for health information*

Our corpus offers several examples of these proof and validation processes, which we have listed in Figure 6.3. Only 23% of the pamphlets base their entire demonstration on a popularized scientific discourse without any references to the work of a doctor, a scientist or a specialized organization, references to experiments or experimental protocols or graphs. This low proportion makes us aware of the relationship between scientific data (raw or formatted) and the reliability or credibility of health information and also attests to a strong emphasis on experimentation and testing.

21 Jeremy Bentham in his *Traité des preuves judiciaires* [BEN 23] theorizes evidence as "a fact assumed to be true, which is taken to serve as a ground of credibility for the existence or non-existence of another fact".

6.3.3. *The figurability of science*

The last aspect we wish to address is what Daniel Jacobi [JAC 05] calls the "figurability" of science. According to him, popularized information relies heavily on images because the reader needs to be able to visualize concepts and information. Jacobi explains that "the quest for figurability" is an attempt to transcribe concepts into images – visual or textual – that are easier to assimilate. Our corpus shows an interesting array of graphs and diagrams that seem to mark the advent of a codified representation of scientific information. Thus, if these graphs are most of the time very simple, even sometimes erroneous, they contribute to the figurability of science and represent an obligatory passage in our corpus.

The turn of the 20th century was marked by great advances in the field of images, both still and animated [DOU 11, p. 14]. At the same time, statistics took on a higher status and became an instrument of proof and even of government [DES 14]. This favorable conjuncture between the accessibility of images and the complexity of statistics justifies in part the success of the graphs or statistical representations, hybrids of images and data, which are found abundantly in our corpus.

The term "image" is used to indicate a visual information or informative image, as opposed to the textual information in our mediums of communication. If we insist from the outset on the informative content of the images or visuals that we study, thus rejecting the hypothesis of a simple function of capture or illustration, it is because our document corpus testifies to a preference for a type of image in particular: the graph.

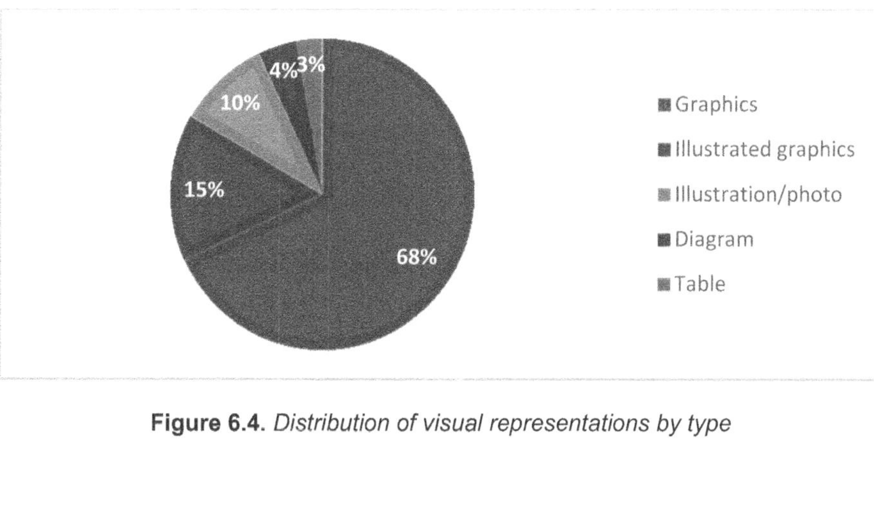

Figure 6.4. *Distribution of visual representations by type*

Figure 6.4 shows the percentage of items for each type of visual element identified in our corpus. Thus, out of a total of 82 pamphlets, more than 60% include a graphic representation (bar graph, area graph, sector graph and combined graph) and about 15% include what we will call here an "illustrated graph", a sort of hybrid between conventional graphic representation and illustration. The images are therefore mostly informative and signal the transition to the documenting mode, understood as an experience of reality produced through the prism of information and not of morality [ODI 11]. As we mentioned when we looked at the regime of expertise and proof in the corpus, most of the graphics "confirm" the truthfulness of the information contained in the titles of the media, and in this respect, they can be qualified as "evidential facts". The visuals in our corpus are thus scientific, if not expert, proof of the assertions that serve as titles for our pamphlets. However, the public is not always equipped to read and understand this type of highly codified formatting that is characteristic of expert discourse [LAV 17, p. 17]. This difficulty put an end to any possibility of contesting the arguments and made this communication space a space for imposing a certain experience of reality, as we will see in the last part.

6.4. Information and influence

> Hardly an aberration, then, Scientific Temperance pioneered a form of lay influence that would extend across our century. As the largest and most successful movement of its kind, it provides a unique window onto the dilemmas and dangers of popular curricular control in an age of expertise [ZIM 99, p. 116].

This quote from historian Jonathan Zimmerman invites us to question the boundary between health information and ideological discourse, and how the appropriation of health information by lay actors can transform the very nature of that information and call into question its objective value. Zimmerman invites us to consider scientific temperance as a fusion between admiring appropriation of scientific expertise and the persistence of a conservative philosophy of sin, individual freedom and personal regeneration. It is precisely this last dimension that hygienists also rejected, since to them the idea of health information tinged with morality was in fact detrimental to the concrete objective of improving the general state of health of the population, starting with the youngest.

Taking up here the idea that the appropriation of health information by a lay public with a very marked political agenda calls into question the informative dimension of our pamphlets in order to emphasize their persuasive dimension, we will show how the visual and textual rhetoric of the images in our corpus participated in the creation of a "*truth effect*" which in reality served an ideological discourse almost as much as an informative one. This further invites a nuanced understanding of the idea according to which the advent of health information in our preventive corpus signals an effective change of approach towards prevention since it seems to be above all a change of "aesthetic" in the broad sense of the term. We seek to bring out the concrete mechanisms of the instrumentalization of health information through several examples from the corpus.

6.4.1. *Handling health information*

The combination of visual and textual rhetoric encourages a misreading of health information in pamphlets. It is not so much the veracity of the information that is questioned as the way in which the public is encouraged to interpret it by the construction of the media and the layout of the information.

The relationship between text and image in pamphlets has the function of a relay [JOL 15]. This function presupposes a complementarity between text and image, where the text fills in the gaps in interpretation left by the image. Where the titles of the pamphlets can be qualified as "general" (general statements on the effects of alcohol), the graphics of the corpus represent "particular" realities (derived from scientific data). It is through the combination of these two scales of relevance that our corpus is set up as a space of health and preventive communication, deriving generalities from particular realities.

In pamphlet 9 (Figure 6.5), entitled, "Parental Alcoholism Diminishes Children's Vitality", the bar graph represents the results of an experiment conducted between 1900 and 1904 to assess the percentage of "weak" and dead children in 50 abstinent and 59 drinking families in a Finnish village (see caption below the title). The particularity of the experiment – far removed from the American context – is here masked by the generality of the title, *a fortiori* if we consider the small size of the sample and the limitations of the experiment. This effect of "generalization" or amplification – depending on whether one insists on the number of individuals concerned or the universality of the message – is observable on numerous occasions in the corpus and participates in an erroneous interpretation of the information through the graphic representation. Analyzing the relationship between writing and communication and specifically particular forms such as tables or lists, Jack Goody stated: "[…] since the table is essentially a graphic process […], it is possible that, by its two-dimensional and fixed character, it simplifies the reality of the oral discourse, to the point of rendering it almost unrecognizable, and that therefore it reduces our comprehension of it instead of increasing it" [GOO 86, p. 111]. These considerations are enlightening for our graphs because they propose a scientific and sociological simplification of reality, likely to correspond almost perfectly to the arguments contained in the Temperants' preventive message. This control over the interpretation of health information was largely favored by the typographic choices made by the Federation, such as the oversized title, the selected color scheme and the font, which was resolutely modern for the time [SEM 16, p. 75].

Figures 6.5 and 6.6. *Pamphlet 9 (STF) and Pamphlet 2 (STF)*

Health information thus masked a persistent moral and conservative discourse under the guise of scientificity. Mary Hunt herself perfectly expressed the ambiguity of the opposition between science and morality. Answering her detractors from the hygienist fringe, she declared that:

> Hygiene without physiology is mere exhortation. [...] It would be as reasonable to propose, in the age of steam and electricity, a return to diligence as it would be to say that moral persuasion should take the place of physiological instruction at Temperance in our public schools [ZIM 99, p. 116].

This quotation proposes an appropriation of the same argument by two opposing sides. Thus, where the hygienists criticized Mary Hunt for giving too much space to physiological and moral justification in her approach to health, the latter turned the same argument against them, implying that inculcating behaviors without explaining their physiological cause was comparable to morality. In the minds of the Temperance movement, health information provided the necessary and sufficient scientific guarantee to justify norms that were ultimately dictated by the morality of the white Protestant middle class from which the members of the Federation and the League came. Health information was a scientific veneer on a moral approach to alcohol and constituted a form of masking of the moral argument via the choice of the documentary mode [ODI 11].

The persistence of the layperson's analysis among that of scientists – strongly marked by morality – can be observed through a type of representation that recurs in our corpus: the illustrated graph (see pamphlet 2: Figure 6.6 (right)). This type of visual, which can be likened to a graph but retains the features of an illustration, reflects the impossible dissociation of morality and health information in the semiotic system of our corpus.

"Illustrated graphics" represent 15% of the visuals in the corpus, so they are the second most frequent type of image after conventional graphs. The visual respects the codes of graphic and statistical representation, but it also borrows from the field of illustrative drawing. A quick iconic analysis shows us that the choice of the representation of children, with characteristic attributes such as the nightgown or the doll, gives the support an emotive dimension that the other types of visuals of the series do not have. The addition of a child's grave on the pamphlet contributes to the *pathos* and reinforces the evocative force of the image. What is then only statistical, or proportional, becomes visualizable and the innocence and fragility of the child, which are highlighted here, come to recall the past preventive tradition, focused on *pathos* and morality. These drawings, *a priori* far from the informative or scientific image, nevertheless respect the codes of the latter by playing here on the proportions or on the figures, so that the images in the childish style come to have a value of proof very similar to that of the conventional graphs. The emotive and moralizing dimension of the illustrations is thus masked in two ways: on the one hand, by borrowing from a semiological system belonging to science and the documenting mode, and on the other hand, by inserting these visuals into a framework identical to that of the graphs, which favors a rapid, and therefore sometimes erroneous, assimilation of the relationship between assertions and evidence.

6.4.2. *Instrumentalizing health information*

The process of medicalization, which began briefly with the irruption of health information in our corpus, is part of very strong ideological currents and represents, in essence, an enterprise of normalization and moral legitimization. The health objective of the prohibitionist campaign is to be considered as a "secondary" objective behind a political, financial and statutory priority for the leaders most invested in the cause.

6.4.2.1. *The limits of medicalization*

The process of medicalization was initiated in our corpus insofar as the pamphlets provided as a foundation a scientific and physiological justification to the effects of alcohol consumption [AIA 98, p. 5]. From then on, and in the absence of treatment, the affirmation of the prophylactic value of total abstention – dear to the Temperants – led to its universal recommendation and marked the movement to the preventive register. It was not so much the demonstration of the universal effectiveness of abstention that led to the desire to have national prohibition adopted, but rather the transformation of practical norms dictated by moral concerns – temperance – into universalizable social norms that found medical justification [AIA 98, p. 7]. The choice to switch to health information is not therefore to be separated from the political struggle of the prohibitionists, but rather invites us to consider our series of pamphlets as tools of active prevention in a society where the real struggle was conducted on the side of passive prevention [RAU 13] with the introduction of national prohibition.

> If, at the cultural level, the medicalization of society corresponds to the reformulation of social problems in medical terms, we can add that, at the political level, it implies the legitimization of this reformulation. This is the meaning of these situations, which are becoming more and more numerous, where politics is expressed in medical language, where health legitimacy is imposed on public action [AIA 98, p. 5].

6.4.2.2. *Instrument of ideological legitimization*

We have previously discussed the idea that the pamphlets mobilizing health information in our corpus could be classified into three broad categories of effects or consequences of alcohol: on individual health, on work and on family members. This structuring of reality and the analysis of the representation of these different domains in the corpus allows us to identify many points in common with the ideological and political dimension of the prohibitionist campaign as described by many historians. Figure 6.7 shows the proportion of pamphlets addressing each type of effect in our two series.

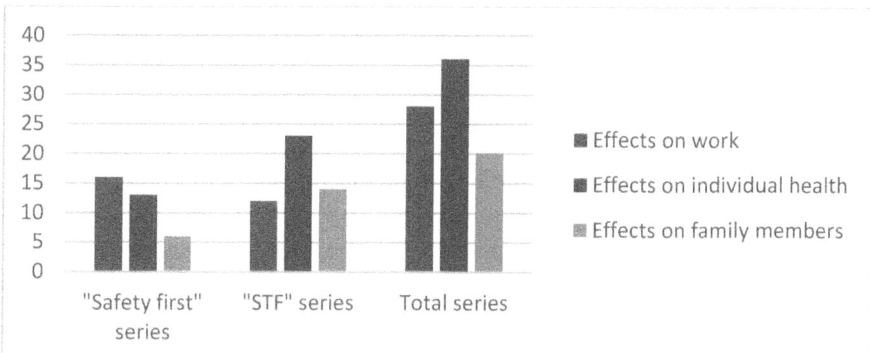

Figure 6.7. *Thematic classification of pamphlets by effect*

It is not surprising to observe that the "Safety First" series, whose objective was the prevention of occupational hazards, dealt first and foremost with the consequences of alcohol on the workplace, and in particular on the quality of work. The "STF" series, on the contrary, did not deal much with the effects of alcohol on the workplace, but focused on the effects of alcohol on individual health and, another striking fact, on family life. However, the graph can be strongly qualified by the fact that the category of effects on individual health tends to approach the effects of alcohol on the body through the prism of performance, and to emphasize the reduction in sensory acuity and the lengthening of recovery time, two factors which could be considered as detrimental to performance in a professional context, much more so than in a personal and family context. This approach to individual health through the prism of performance invites us to qualify the very idea of "individual health" and to consider it as a "pawn" in an economic, political and statutory ideology.

Economic and political, because for some historians the campaign to reform drinking habits was in the hands of men of power and wealthy capitalists who saw in prohibition a means of establishing a new industrial order, and therefore an opportunity for profit [RUM 89, p. 1]. For these men of power, support for national prohibition was an expression of a "utopian vision of society at the heart of which was an abstinent and cooperative workforce whose constant occupation was to produce wealth for others" [RUM 89, p. 1]. The pamphlets we analyze and the "Safety First" series in particular directly support this historiographical approach by equating alcohol prevention with improved worker performance. Abstinence was then assimilated to a real lifestyle, promoting values such as industry or moderation and illustrating perfectly the capitalist attempt to transform the social, cultural and psychological landscape of the country in order to serve the interests of the rich and influential recruiters.

Statutory also because, for the historian Joseph Gusfield, on the contrary, prohibition, temperance and abstinence are to be considered as symbols of the "symbolic and statutory" supremacy of the Protestant and rural middle class, which was able to assert itself in a context of industrialization, urban development and Catholic immigration [GUS 86 p. 99]. The Temperance movement, mostly evangelical, defined their ideal of society according to conservative and moral principles strongly marked by Victorian influence, and attached importance to family values and traditional gender roles [GUS 86, p. 4]. In this sense, the division by types of effects of the pamphlets mobilizing health information in our corpus is to be interpreted as a reflection of the basic values of this conservative and liberal fringe of society concerned with the control of rampant industrialization, the defense of the family unit and the good birth rate, and finally of a healthy way of life devoid of the alcoholic excesses traditionally associated with the immigrant fringe of the population.

6.5. Conclusion

> No candid mind will deny that knowledge does not always preserve from evil, but the human mind is so made that it is disposed to be convinced by facts, and to regulate its action accordingly, otherwise why should we burden ourselves with education? [ZIM 99, p. 122].

These words by Cora Frances Stoddard, inspired by the highly respected Mary Hunt, attest to the importance that the Temperants placed on health information and its transmission. The social, domestic, industrial and economic consequences of alcohol were, for them, to be read primarily through the prism of alcohol's physiological effects. In fact, only health information, because it favored a personal and not a community appropriation of prohibitionist arguments, was likely to really convince the public.

Far from being limited by its potential for personal appropriation and social acceptability, we have seen in this chapter that health information is to be considered as an issue on several levels. Health information was a preventive tool whose argumentative potential was appealing to an increasingly wide range of actors. It was therefore a condition for the possibility of reforming individual and community health behaviors. Health information is also to be grasped as a field of competence and influence for actors coming from civil society who found, through the acquisition of an expertise in health, a form of legitimacy to publicize the issue of alcohol. Finally, the omnipotence of health information must be qualified, because in our corpus, it is similar to an aesthetic artifice that only clumsily masks the persistence of a conservative ideal. Centered around moral and religious values and serving the economic and statutory interest of the powerful, the universal dimension of health as a domain of bodies was here subordinated to a system of values modeled on the drinking practices of the public. The example of scientific temperance is thus in line with the work on medicalization, which designates this process not as the application of an imminent medical or scientific principle but as a new form of social and moral normalization borrowing from the medical register to establish its legitimacy.

In conclusion, it should be noted that the Scientific Federation for Temperance, so devoted to the defense of health information, was quickly obliged to favor a socio-demographic and statistical approach to alcohol at the expense of the physiological approach. Easier to capture on a communication medium than fatigue or dementia, crime, poverty and violence were, for the "general public", familiar and popular subjects in the communication media. The home setting was much easier to picture than the inner workings of the human body, and health information gradually disappeared from Federation publications in the early 1920s, only to return much later.

6.6. References

The pamphlets used in this chapter are all from the archives of the Anti-League Saloon, Westerville, Ohio.

[AIA 98] AIACH P., DELANOË D., *L'ère de la médicalisation – Ecce homo sanitas*, Anthropos, Paris, 1998.

[BAN 09] BANCROFT A., *Drugs, Intoxication and Society*, Polity, Cambridge, 2009.

[BEA 85] BEAUD P., Médias, médiations et médiateurs dans la société industrielle, PhD thesis, Université Stendhal, Grenoble, 1985.

[BEN 23] BENTHAM J., *Traité des preuves judiciaires*, Bossange Frères, Paris, 1823.

[BER 07] BERLIVET L., "Les démographes et l'alcoolisme. Du 'fléau social' au 'risque de santé'", *Vingtième siècle – Revue d'histoire*, vol. 95, no. 3, pp. 93–113, 2007.

[BER 08] BERGERON H., "Qualifier en politique: L'exemple du problème alcool", *Santé publique*, vol. 20, no. 4, pp. 341–352, 2008.

[BER 16] BERRIAU N., "Les données de santé, une opportunité pour les professionnels de l'information", *I2D – Information, données & documents*, vol. 53, no. 3, pp. 30–31, 2016.

[BOU 04] BOURRET C., "Réseaux de santé et nouveaux métiers de l'information", *Documentaliste-Sciences de l'information*, vol. 41, no. 3, pp. 174–181, 2004.

[CEC 08] CECCHI C., "La place de l'information dans la décision en santé publique", *Santé publique*, vol. 20, no. 4, pp. 387–394, 2008.

[CHA 97] CHARAUDEAU P., *Le discours d'information médiatique: La construction du miroir social*, Nathan, Bry-sur-Marne, 1997.

[CHO 00] CHOO C.-W., *Information Management for the Intelligent Organization: The Art of Scanning the Environment*, ASIS Monograph Series, USA, 2000.

[COK 07] COKER J., *Liquor in the Land of the Lost Cause: Southern White Evangelicals and the Prohibition Movement (Religion in the South)*, University Press of Kentucky, Lexington, 2007.

[COM 88] COMBETTES B., TOMASSONE R., *Le texte informatif: Aspects linguistiques*, De Boeck, Brussels, 1988.

[DAR 06] DARGELOS B., "Genèse d'un problème social. Entre moralisation et médicalisation: La lutte antialcoolique en France (1850–1915)", *Lien social et Politiques*, no. 55, 2006.

[DEO 14] DE OLIVEIRA J.-P., "La communication publique à l'appui d'une stratégie de repositionnement de l'Etat dans l'espace public", *Questions de communication*, vol. 26, no. 2, pp. 235–255, 2014.

[DES 14] DESROSIERES A., *Prouver et gouverner*, La Découverte, Paris, 2014.

[DOU 11] DOUGUET F., FILLAUT T., SCHWEYER F., *Image et santé, matériaux, outils, usages*, Presses de l'EHESS, Paris, 2011.

[EME 77] EMERSON R., MESSINGER S., "The micro-politics of trouble", *Social Problems*, vol. 25, no. 2, pp. 121–134, 1977.

[FAV 05] FAVEREAU E., "Le journalisme, de l'information médicale à l'information de santé", *Les Tribunes de la santé*, vol. 9, no. 4, pp. 21–26, 2005.

[GOO 86] GOODY J., *La raison graphique*, Editions de Minuit, Paris, 1986.

[GUS 81] GUSFIELD J., *The Culture of Public Problems*, Chicago University Press, Chicago, 1981.

[GUS 86] GUSFIELD J., *Symbolic Crusade: Status Politics and the American Temperance Movement,* University of Illinois Press, Urbana & Chicago, 1986.

[JAC 05] JACOBI D., *Les sciences communiquées aux enfants: Travail d'édition et éducation non formelle*, PUG, Grenoble, 2005.

[JOL 15] JOLY M., *Introduction à l'analyse de l'image*, Armand Colin, Paris, 2015.

[KER 85] KERR K.A., *Organized for Prohibition: A New History of the Anti-Saloon League*, Yale University Press, New Haven, 1985.

[KIV 14] KIVITS J., HANIQUE M., JACQUES B. *et al.*, "L'appropriation de l'information médiatique au sujet de la prévention et du dépistage des cancers", *Le Temps des médias*, vol. 23, pp. 151–163, 2014.

[LAV 17] LAVOIE V., *La preuve par l'image*, Presses de l'Université du Québec, Quebec, 2017.

[LEG 97] LEGLISE I., GARRIC N. (eds), *Le discours d'information médiatique: La construction du miroir social*, Nathan, Bry-sur-Marne, 1997.

[MAR 08] MARTIN S.C., *Devil of the Domestic Sphere: Temperance, Gender, and Middle-Class Ideology, 1800–1860,* Northern Illinois University Press, DeKalb, 2008.

[ODI 11] ODIN R., *Les espaces de communication*, PUG, Grenoble, 2011.

[OLL 13] OLLIVIER-YANIV C., "Communication, prévention et action publique: Proposition d'un modèle intégratif et configurationnel", *Communication & Langages*, vol. 176, no. 2, pp. 93–111, 2013.

[PAI 05] PAILLIART I. (ed.), *La publicisation de la science: Exposer, communiquer, débattre, publier, vulgariser hommage à Jean Caune*, PUG, Grenoble, 2005.

[PER 09] PERETTI-WATTEL P., MOATTI J.-P., *Le principe de prévention: Le culte de la santé et ses dérives*, Le Seuil, Paris, 2009.

[POU 10] POURCHET M. (ed.), *Le cancer dans les médias 1980–2007*, Centre d'étude des discours, images, textes, écrits, communications and Institut national du cancer, Boulogne-Billancourt, 2010.

[RAU 13] RAUDE J., "Les stratégies et les discours de prévention en santé publique: Paradigmes et évolutions", *Communication & Langages*, vol. 176, no. 2, pp. 79–92, 2013.

[ROM 15] ROMEYER H., "Le bien-être en normes: Les programmes nationaux nutrition santé", *Questions de communication*, no. 27, pp. 41–46, 2015.

[RUM 89] RUMBARGER J., *Profits, Power and Prohibition: Alcohol Reform and the Industrializing of America, 1800–1930*, State University of New York Press, Albany, 1989.

[SCH 09] SCHWEYER F., "Politiques publiques: L'impasse d'une conception balistique", *Après-demain*, no. 10, pp. 29–33, 2009.

[SEM 16] SEMPRINI A., *Communiquer par l'image: Trois essais de culture visuelle*, PULIM, Limoges, 2016.

[ZIM 99] ZIMMERMAN J., *Distilling Democracy: Alcohol Education in America's Public Schools 1880–1925*, University Press of Kansas, Lawrence, 1999.

7

Professional Uses of Specialized Information in Ophthalmology. Interview with Pascal Four, Doctor of Ophthalmology

Pascal Four is a doctor of ophthalmology and works in private practice in Romans, France. He has created a documentary database and testifies about his information practices in relation to the exercise of his profession.

7.1. Training and areas of intervention in ophthalmology

After beginning his medical studies in 1984 at the Claude Bernard Lyon 1 University, Pascal Four completed the first part of his training in general medicine, then passed the internship exam in 1991 in the surgical field and chose the ophthalmology specialty after having completed a few semesters. Then, he defended his thesis in 1995 under the direction of Luc Durand in the specialty of ophthalmology on "*The refractive defect after radial keratotomy and its treatment. Based on 1069 successive interventions*". He settled in Romans in 1997 and has now 23 years of experience in private practice. During his internship, he also practiced in a hospital. Then, in parallel to his private practice, he performed occasional surgeries at the hospitals of Romans and Valence.

Interview by Viviane CLAVIER, conducted on July 25, 2020.

Dr. Four practices general ophthalmology, including glasses and contact lenses. He provides screening and treatment for glaucoma, AMD[1] and diabetic retinopathy. He treats eyelid pathologies (inflammatory, infectious, tumoral). He monitors many diabetic patients for diabetes-related retinal damage. He also monitors chronic inflammatory diseases of the ocular surface (dry eye, allergies or other rarer surface diseases). He manages emergencies: foreign bodies, trauma, conjunctivitis, inflammatory diseases of the eyelids, socket or eyeball (such as uveitis or rare diseases), lacrimal obstruction, retinal detachment and other acute retinal diseases, optic neuritis and retinal vascular accidents. He monitors ophthalmological signs that appear in general diseases and is sometimes asked by general practitioners to perform ophthalmological check-ups, for example, in the case of a disease of the central nervous system (aneurysms, MS[2], etc.). He also treats strabismus and ocular motor disorders. Having stopped operating two years ago, Dr. Four now diagnoses and refers patients who require surgery such as cataracts, palpebral tumors, retinal detachments and somewhat rarer diseases of the macula that may require surgery.

7.2. Information useful in the context of medical practice

Dr. Four often needs information, especially on rare diseases, on the modalities of diagnosis and treatment and on possible correspondents specialized in this type of pathology. He refers to colleagues who are particularly interested in these rare diseases and who regularly communicate on these subjects. When the pathologies are complex, it is also interesting to have the opinion of a more specialized correspondent. Patients are then referred to this correspondent who then gives his opinion. It is therefore necessary to have a list of specialists and to keep them in mind during the consultation. As ophthalmology is increasingly subdivided into specialties, this list must be updated regularly. Dr. Four indicates that there are currently specialists in glaucoma, strabismus, eyelids, congenital diseases, medical retinal diseases, surgical retinal diseases, ocular surface diseases, uveitis, ocular tumors, refractive surgery, cosmetic surgery of the eyelids, orbital surgery and neuro-ophthalmology among others. Sometimes, within these specialties, there are even ophthalmologists who are more particularly competent in this or that disease. It is therefore useful to have read articles to

1 Age-related macular degeneration (AMD).

2 Multiple sclerosis (MS).

identify a network of professionals who will be able to take care of the patients.

The notion of patient care refers to the way in which diagnostic assessments are established and to care protocols, which may vary from one author or from one period to another. Dr. Four needs information about certain pathologies whose care is constantly changing and sometimes controversial. The academic societies and ophthalmologists who are authorities in this or that pathology regularly issue new recommendations that significantly vary over the years (e.g. in TIA[3], monitoring of patients treated with Plaquenil, uveitis or optic neuritis assessments, etc.).

Whatever the information, Dr. Four often needs it during the consultation, hence the interest of his document base which allows him to very quickly obtain the elements useful for the diagnosis, the recommendations concerning the management of certain complex pathologies as well as the data allowing him to inform the patient in accordance with the current state of medicine. According to Dr. Four, it is impossible to "retain everything" because knowledge is so extensive. There is indeed "encyclopedic" data on strabismus, uveitis, glaucoma, retinal diseases or general diseases that can affect the eye. It is even more difficult to "know everything" when the treatment is constantly evolving or when it concerns infrequent pathologies that are encountered only once or twice during the career of an ophthalmologist. For example, Dr. Four was recently confronted with Brown's syndrome (congenital paralysis of an oculomotor muscle, very rare, about one case in 20,000 births), a pathology that he knows but has never encountered in over 20 years of practice. In this case, it is very useful to have a "reminder" to summarize all the diagnostic elements and therapeutic possibilities. Quick access to information is an asset.

Finally, the need for information is related to medicines. On this point, there is a database specific to medicine, the Vidal dictionary, as well as publications that deal with treatments and how to use drugs. There are sometimes different indications between these two categories of sources. For example, Skiacol is a medication used to dilate the pupil of children before refraction. In the Vidal dictionary, only one drop is recommended, possibly a second drop if dilation is insufficient, whereas all specialists in ocular refraction in children recommend three drops to systematically block

3 Adult transient ischemic attack (TIA).

accommodation sufficiently. It is therefore essential to rely on articles in the event of a medico-legal problem and to argue that the generalized practice of administering three drops is validated by the scientific community. In addition to the problems of variable indications, there is also the question of the use of drugs in compliance with the MA[4] or outside the MA. This MA, available in the Vidal dictionary, defines for each drug the precise context in which it can be used (medical indication, dosage and route of administration). However, the use of drugs outside the MA is frequent because of their effectiveness in certain unlisted pathologies. For example, Diamox, which is usually used to treat intracerebral edema, is often used off-label for macular edema, since alternative treatments involve greater risks for the patient. In doing so, the physician incurs medical and legal liability by prescribing this treatment outside the MA. It is therefore important in such situations to rely on the recommendations of academic societies or colleagues who are authorities in a given field of ophthalmology.

7.3. Access to knowledge and information sources

Access to information in the field of ophthalmology is complex, not because of the scarcity of publications, since there are many articles, but because of the absence of recent summaries that take stock of what has been written on a subject. When a subject is controversial and several authors express divergent opinions, as is the case for retinal vein occlusion, for example, it is the physician's responsibility to summarize the information in order to know how to proceed. These summaries can sometimes be found in books, but they are not updated. Although there are recent publications, it is difficult to form an opinion in the absence of an overview of the articles that deal with a given problem.

Knowledge in the field of ophthalmology does not all evolve at the same speed. For example, the treatment of cataracts is well codified and is not currently evolving very much, as indicated by Dr. Four, who has spent his entire career using the same type of phacoemulsification procedure, the current reference technique for cataract surgery, and which has endured despite all other attempts to dethrone it (laser, femtosecond laser, etc.). In this case, despite an abundance of sources, there is a consensus unlike some areas, which are more subject to evolution and controversy within the

4 Marketing authorization (MA).

ophthalmological community, such as TIA or uveitis. On the other hand, there is a lack of information on rare diseases. However, Dr. Four notes that since the beginning of his career, many rare pathologies have been discovered. But in this field too, he regrets the absence of summaries, which would make it possible to fill the information deficit related to the scarcity of sources, the fragmented nature of knowledge which is explained by the few authors who study these pathologies and also the few patients concerned by these pathologies (certain diseases are described from only 10 cases). This situation makes it difficult to find articles and precise information and, although there is little chance of encountering these pathologies in our career, being able to diagnose them remains a necessity.

The sources of information consulted are multiple:

– the Vidal online dictionary;

– monthly professional journals that publish articles in French[5]: *RéfleXions Ophtalmologiques*, a continuing education journal that has been published since 1996[6]; *Les Cahiers d'Ophtalmologie*[7], a journal published for 16 years by a city ophthalmologist, oriented toward daily medical practice and which "gives a place to material and includes a News section"; *Le Journal Français d'Ophtalmologie* published by the *Société Française d'Ophtalmologie*[8], which provides "reliable scientific information and offers numerous tools for continuing medical education (thematic files, round tables, surgical techniques, images selected for their visual quality, anatomical and clinical charts)"; *Réalités ophtalmologiques*[9] published by a press and medical communication group called *Performances Médicales*, which has been a limited company since 1990. This group publishes a collection of medical journals, including a monthly ophthalmology journal; and finally, the *Revue de l'Ophtalmologie Française*[10], a bi-monthly journal

5 With the exception of the *Journal Français d'Ophtalmologie* which publishes half of its articles in English.

6 https://www.jbhsante.com/specialites/ophtalmologie/reflexions-ophtalmologiques (accessed July 29, 2020).

7 https://www.cahiers-ophtalmologie.fr/les-cahiers-en-bref (accessed July 29, 2020).

8 https://www.sfo.asso.fr/content/journal-fran%C3%A7ais-d%E2%80%99ophtalmologie (accessed July 29, 2020).

9 https://www.realites-ophtalmologiques.com/le-groupe/ (accessed July 29, 2020).

10 https://www.speps.pro/publication/revue-de-lophtalmologie-francaise/ (accessed July 29, 2020).

published by the *Syndicat National des Ophtalmologistes de France* (SNOF);

– the information published by the academic society SFO (*Société Française d'Ophtalmologie*): its website, its reports;

– Elsevier's[11] *EM Consult* database for private practitioners, which includes more than 570,000 validated articles, 120 medical, paramedical and scientific journals and 47 treatises;

– notes taken at discipline conferences or web conferences;

– specialized works, but whose data are often old;

– Google search for interesting and up-to-date information such as the American Academy of Ophthalmology (AAO) website or Eyewiki.

A large number of these sources are paid for and are subject to subscription: *Le Journal Français d'Ophtalmologie*, *EM Consulte*, subscription to the SFO website, SFO reports, participation in congresses, specialized books. The rest is free and includes advertising related to ophthalmology. Some subscriptions are very expensive, such as *EM Consulte*, whose search interface remains complex to use. Dr. Four regularly consults these journals to produce summaries for his document base and then has quick access to the information.

7.4. Time spent searching for information, criteria for evaluating information

Dr. Four spends a lot of time searching for information. Although the time spent on this activity is not constant throughout the year, he estimates it to be 15 hours per week to process articles and bibliographies. This activity is considered useful and allows him to be "sharper" professionally. Searching for information on rare diseases is the "most time-consuming" because of the rarity of publications, which requires spending a lot of time finding quality sources. In addition, research on controversial diseases is also time-consuming, as they are constantly evolving and there is no clear summary, which requires him to consult numerous articles on a regular basis to keep up with the latest news and form an opinion.

11 https://www.elsevier.com/fr-fr/clinical-solutions/emconsulte (accessed July 29, 2020).

In assessing the reliability of the sources and the quality of the information, Dr. Four first looks for consistency of content with his knowledge of the subject – if in doubt, he compares the articles with other similar publications – and for accuracy of data, which indicates good quality work and a rigorous approach. Secondly, the author and date are taken into account, with greater credit given to recent articles, as well as the quality of the bibliography. Finally, the possibility of a conflict of interest with the pharmaceutical industry is taken into consideration, which may constitute a bias in the writing of the article, particularly in the issuing of opinions. For some highly specialized surgeons, we must also consider the possibility of withholding information and knowledge in order to maintain an unshared place of excellence.

7.5. Information exchanged with patients

Generally speaking, patients often need to know and understand the disease they have, so the doctor must be able to provide them with clear information, in line with the current medical knowledge and adapted to the patient's ability to understand. The information will not be the same for a 40-year-old executive as for an elderly person with Alzheimer's disease. Apart from any pathology, patients are eager to ask about preventive measures to avoid problems (diet, hygiene, sun protection, precautions with screens, etc.). These recommendations evolve over time and are sometimes subject to controversy. This is the case for food supplements proposed for AMD, for example. An updated summary of knowledge is therefore useful in this case. In the context of an ocular disease, Dr. Four is always surprised by the need for patients to find a cause (external agent, diet, medication, behavior, etc.). For example, presbyopia is often associated with the use of screens, when in fact it is due to the physiological aging of the lens. The development mechanism of a disease (physiopathology) is not the first question of the patients who seek more to identify an external cause at the origin of their problem. In many cases, however, the cause is not known.

In the particular case of a rare disease, patients may have already taken advice prior to the consultation, or searched for information on the Internet. They may therefore have very specific questions and the document base is very useful to answer their questions.

However, not everyone seeks information. Dr. Four estimates that about 70% of his patients ask for information and 30% do not. It is true that when patients consult for a renewal of their optical correction, there is often little questioning on their part, except in the case of children whose parents wish to know the cause of the refractive defect and the potential evolution.

When patients come to the doctor's office, they often have information, the youngest having found it on the Internet, the oldest from their family and friends. Dr. Four sees the patients for the first time and proposes, in certain cases, a complementary assessment when a pathology is suspected (glaucoma, retinal degeneration, AMD, optic neuropathy, etc.). The information is voluntarily limited during the first visit in order not to worry the patients unnecessarily. The suspected diagnosis may or may not be mentioned depending on its severity and the patient's degree of anxiety. An explanation may be given to justify the performance of additional tests. Between the two consultations, patients tend to seek information on their own. They return to the second visit with more questions as well as with more knowledge that sometimes allows them to better understand the doctor's explanations.

Today, there are many sources of information on ophthalmology available to the general public. For example, some ophthalmology centers have developed information sites for patients: "Why have myopia surgery? What are the techniques? etc.". This is the case for cataract surgery, refractive surgery, cosmetic surgery of the eyelids, etc. In addition, certain hospital services also provide information on ophthalmology via their website: the Quinze-Vingts Hospital in Paris or the Adolphe Rothschild Ophthalmological Foundation in Paris. Dr. Four also consults these sites to assess the discourse on a given pathology and to verify the level of information provided to patients. According to Dr. Four, outside of benign pathologies, excessive information often leads to increased anxiety and a greater need for information. According to him, "it is important to find the right balance in what is said to the patient and to adapt it to his or her psychology".

The information is mainly transmitted orally during the consultation. Standard information sheets exist for certain frequent pathologies. Some information sheets are mandatory from a medico-legal point of view when surgery is envisaged. The patient must then sign an informed consent form. In ophthalmology in France, these documents are written by the SNOF (*Syndicat National des Ophtalmologistes de France*) and by the SFO (*Société Française d'Ophtalmologie*).

7.6. Knowledge organization and structuring of the document base

Since 1997, Dr. Four has been writing abstracts from journal articles, books and conferences he attends every year. The journals he has consulted over the past 20 years now occupy about 10 linear meters. At the beginning of his career, as soon as he read an article, he took notes and archived them by theme in an Access database. These notes, which took the form of comments and more or less developed summaries, were recorded in documents of varying size, from a few lines to three or four pages. However, access to this database was complex and did not allow for an overall view, as the classification was brief and the summaries were very numerous. Between late 2018 and May 2020, Dr. Four reduced the 16,000 pages of personal summaries to 3,623 pages summarized and categorized by topic and pathology and saved them in the Google Docs format available on Google Drive. When a pathology fell under several themes, hyperlinks were established between the themes. According to Dr. Four, the referencing of the articles does not follow academic standards and is not very usable by others: journal numbers were entered (but not journal names), as well as article titles, primary author, date and pages. For each publication, Dr. Four assigns a score from 1 to 5 based on the interest of the article. Some abstracts include links to external sources or professional information sites; some articles are uploaded to Google Drive in pdf format and complement the abstracts to which they are attached with a hyperlink. The archiving of these documents follows a logical organization in five levels identifiable by a particular typography and a color code.

– On the first level, a table of contents gives access to the summaries gathered in chapters and subchapters. The table of contents covers a dozen themes as shown in Box 7.1.

General; clinical examination; visual function; cornea/conjunctivitis; uveitis; retina/ choroid/vitreous; glaucoma/HPIO; tears/VL/eyelids; orbit; motricity (except POM)/ VB/strabismus/amblyopia; neuro-ophthalmology; general diseases/OPH signs; complementary exams; miscellaneous; therapeutics

Box 7.1. *Level 1 titles*

– At the second level, subdivisions into subchapters are sometimes included. Level 2 titles are not systematic. For example, the level 1 theme “Retina/Choroid/Vitreous” has nine subdivisions corresponding to different pathologies.

Semiology/ anatomy/ congenital disease; vitreous/ tumors; vascular pathology; heredo-degeneration; AMD/ maculopathies; CRSC/ ERD/ rare diseases; myopia/ peripheral/ DR; retinopathy/ DR/ others; miscellaneous/ surgery/ IVT

Box 7.2. *Level 2 titles*

– The third level, contrary to the second, is systematically described and is sometimes presented as the specific terms of a classification. To take the title of level 2 “AMD/maculopathies”, there are five subdivisions corresponding to the types of AMD identified by a numbering, a code corresponding to the file number on Google Docs and a number of pages.

AMD 1 (Ia07a) (50p); AMD 2 (Ia07b) (40p); AMD 3 (Ia07c) (35p); AMD 4 (Ia08) (20p); maculopathies (except AMD) (Ia09) (42p)

Box 7.3. *Level 3 titles*

Note that other modes of organization exist. Level 1 “Cornea/Conjunctivitis” has 14 level 3 subdivisions that refer to the histological description of the cornea, various malformations, tumors, various types of infections, various trophic or degenerative disorders, allergies, etc.

– The fourth level is presented in the form of keywords reduced to a single term (common or proper noun), a multi-term or an acronym. For example, level 3 of “Rare diseases/choroidal pathologies” contains a list of 14 terms in the singular or plural. This organization of knowledge is similar to a terminology list.

White spot syndrome; EEP/plaque maculopathy; serpiginous choroiditis/ AMPPE; CMF/PIC; MEWDS; blind spot enlargement syndrome; acute macular neuroretinopathy; AZOOR; ARPE; angioid streaks; choroidal folds; PEVAC; pachychoroid/choroidal thinning; peripapillary pachychoroid

Box 7.4. *Level 4 titles*

– The fifth level presents Dr. Four's personal summaries, any diagrams, comments and formatted notes.

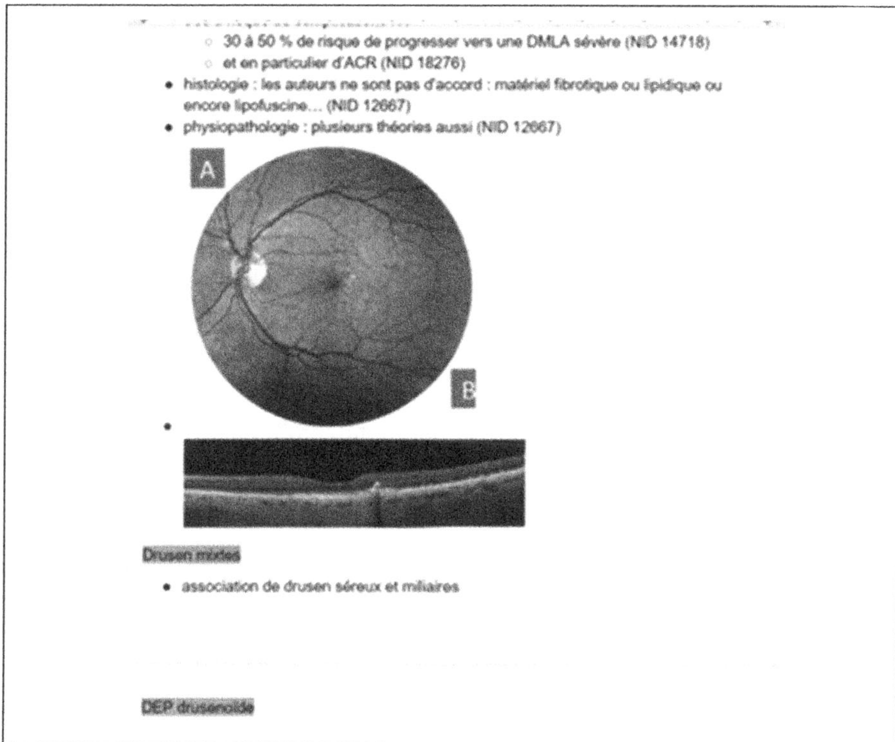

- 30 à 50 % de risque de progresser vers une DMLA sévère (NID 14718)
- et en particulier d'ACR (NID 18276)
- histologie : les auteurs ne sont pas d'accord : matériel fibrotique ou lipidique ou encore lipofuscine… (NID 12667)
- physiopathologie : plusieurs théories aussi (NID 12667)

Drusen mixtes

- association de drusen séreux et miliaires

DEP drusenoïde

Box 7.5. *Level 5 summaries. For a color version of this box, see www.iste.co.uk/paganelli/information.zip*

To conclude this presentation, the organization of knowledge that presides over the structuring of this document base reflects as closely as possible Dr. Four's medical practice. Various contextual factors explain the over- or under-representation of a given theme, such as the age of his patients, which justifies the presence of a highly developed section on age-related macular degeneration (AMD), as well as the editorial context of publications, which feeds the information monitoring. The document database is updated, the oldest abstracts are deleted, except when the documents allow us to justify the management of patients that took place in the past and justify certain therapeutic choices from a medico-legal point of view. The very synthetic formatting of the summaries, which follows a quasi-standardized presentation in the form of lists and sub-lists, succinct and personal comments, and hyperlinks, is designed for rapid access during the consultation and serves as a "checklist" for drawing up an assessment, a protocol or a prescription. In addition, the availability of documents on Google Drive allows him to access the database from his phone from any location and meets the following needs:

– diagnostic assistance;

– therapeutic decision-making;

– possible referral of patients;

– the ability to properly inform patients;

– the justification of a therapeutic decision from a medico-legal point of view;

– communication of information to non-ophthalmologists.

This organization of knowledge is not only inspired by medical experience, it also borrows from other ways of organizing knowledge, such as medical classifications of diseases that are codified and present a wide variety of signs, symptoms or listed conditions. However, in addition to diseases, it also considers anatomy as well as functional and physiological aspects. Although there are subdivisions and codes, the coding is not backward thinking as is the case with the Dewey Decimal Classification, which has a systematicity in the divisions. It also has points in common with medical terminologies, although they do not give the meaning of terms, as is the case with glossaries in ophthalmology. The absence of controlled or free indexing would appear to be a limitation for consultation by a third party, but does not seem to affect the search for information, as the database is familiar to its creator, who uses it daily and is therefore familiar with the keywords that he himself has defined. The summaries are reminiscent of lecture sheets that are intended to memorize courses or to draw up a state of the art for a publication.

Thus, the organization of knowledge, the structuring of the document base and the scriptural form of the documents are designed to accompany medical practice as closely as possible, while taking into account the evolution of knowledge in ophthalmology. Knowledge is evolving very rapidly and refers to anatomy, physiology (of the eye, the socket, the central nervous system), as well as to local, neuro-ophthalmological and general pathology (ophthalmological repercussions of systemic diseases). This knowledge also concerns complementary examinations as well as imaging, which is increasingly present in ophthalmology. We currently speak of "multimodal imaging" which combines several techniques for collecting and processing images. The knowledge encompasses management (diagnostic assessments, medical treatments, laser interventions, surgical techniques) as well as the updating of a professional network of practitioners in various specialties of ophthalmology.

Finally, the knowledge relates to the evolution of knowledge on rare pathologies and the follow-up of scientific controversies within the ophthalmological community. The structuring of the document base, organized in different levels of access to information, is reminiscent of ophthalmology treatises structured in chapters and subchapters with a table of contents. These treatises or introductory works in ophthalmology are intended for students of general medicine. The medical–surgical encyclopedia from Elsevier also has a similar content.

The interview with Dr. Four shows that informational practices in the context of professional practice can in no way be dissociated from questions of research, selection and validation of information. Although this information is essential to medical practice, it is unusable without treatment. Thus, the organization of knowledge in ophthalmology in the form of an updated documentary base responds to individual needs for updating knowledge and memorization, as well as for inclusion in a network of recognized and scientifically active specialists. Paradoxically, while these time-consuming informational activities are considered to be invisible, because they are not counted as part of working time, and because they are carried out outside of the practice, most often at home, it appears that they ensure a certain visibility for this specialist: within the ophthalmological scientific community by participating in scientific events and by updating knowledge in professional journals; within the professional network by regular exchanges with colleagues in specialized fields of knowledge; with patients who expect precise information on the pathologies that affect them.

One of Dr. Four's questions is whether such a database would be useful to all French ophthalmologists and could serve as a national reference for medical practice. Without a doubt, the answer is yes. However, this implementation does not lie in a simple change of scale. Indeed, the contents of the database are designed according to the specificities of Dr. Four's practice context (the clientele, the locality, the way of practicing medicine and exploiting information, the way of reading, of writing abstracts, of selecting and evaluating articles, the writing format, the choice of themes, the codification, etc.). As a result, the organization and prioritization of knowledge reflect such specificities. On the other hand, the principle of providing access to up-to-date medical knowledge validated by specialists via a document base on ophthalmology is in keeping with the library and encyclopedic tradition of shared universal knowledge, an ancestral perspective that still feeds projects for knowledge management systems.

7.7. Appendix: interview grid

Professional activity

1) What activity?

2) Since when? Years of experience?

3) In what context? Liberal?

Documentary practices and the place of information

4) In what professional situations do you feel the need to search for medical information (e.g. search for a scientific article, search a database, ask a colleague a question)?

5) What type of information do you most often need for your activities? Do you find it difficult to find or access certain information?

6) What sources of information (print, databases, search engines, libraries, social media, colleagues) do you typically consult?

7) How do you assess the reliability and quality of information from these sources?

8) How do you feel about the amount of time you spend searching for information on a daily basis? Which tasks related to the use of information are particularly time-consuming?

9) What type of information do your patients often need? How do you feel about the way your patients get information? How do you communicate with your patients?

Database[12]

10) What content? Types of documents, number of documents?

11) What description? Which fields?

12) Indexing with a particular language? Free indexing?

13) Which tool? Why this choice?

14) How is it powered?

15) How often? Since when?

16) Why this database?

17) When do you use it? For what situations? To respond to patients? To colleagues? In very specific cases?

18) Examples of searches in this database? To do what? Examples of use?

19) Sharing the database? With whom? Otherwise why?

20) How does this database complement other sources of information?

12 The concept of a database was not retained in the minutes.

8

Organization of Medical Knowledge: Documentation Techniques Applied to a Macro-domain Underpinned by Socio-political Issues

8.1. Introduction

This chapter evaluates the dynamics and systems of medical knowledge organization developed by information and documentation specialists and explains the specific trends and developments in this field since the 1960s. Knowledge organization is a field of research that brings together researchers from disciplines such as information and documentation science, linguistics, philosophy and computer science [GNO 20, ZAC 10]. After clarifying the key concepts and the methodology of the study, we proceed with a literature review, combined with an analysis of a corpus of knowledge organization systems (KOS) in medicine. An assessment, highlighting the current direction and explaining the positioning of the discipline, concludes the reflection.

8.2. Documentation and Library and Information Sciences and the organization of knowledge

Documentation and Library and Information Sciences form an interdisciplinary field of research, bringing together librarianship and

---

Chapter written by Marcin TRZMIELEWSKI and Claudio GNOLI.

information science, which lies between the humanities and social sciences (HSS) and computer science [HJØ 18]. Furthermore, the current trend is to use the terms Library and Information Science and Information Science as synonyms [HJØ 18]. Reflections on the theoretical and practical notions developed by specialists in this field – i.e. teacher-researchers and professionals of information–documentation (librarians, documentalists, cataloguers, taxonomists, etc.) – are concentrated around "access to knowledge and culture" and "recorded knowledge" [MET 06, p. 43]. One of the fundamental concepts and one of the principal missions of the sciences of information and documentation is to study and consider *document mediation*. As Gérard Régimbeau points out, "document mediation is a mediation of knowledge" [RÉG 11, p. 108]. It manifests itself in the interactions between the information professional and the user of an information system and "requires the implementation of a communication system adapted to the user" [*ibid.*, p. 108–109]. The challenge of this process is to "allow the user to find the information [they] need" so that they "build [their] own knowledge in order to develop [their] own knowledge-base" [*ibid.*, pp. 106–109]. The interactions between the producer of the information, the mediator and the user are implemented within the *documentation devices*, which are "instrumental components" within the information systems [MET 06, p. 54], responsible for the diffusion of the knowledge between these actors, with the aim of the appropriation of the information by the user as content allowing to act [COU 11]. The devices have three dimensions: technical, socio-organizational and socio-economic, which are inextricably linked and interdependent [MET 06]. To enable access, consultation and management of information in information systems, and to implement mediation within documentation devices, professionals, users and computer programs resort to *knowledge organization* [COU 11, GNO 20, HJØ 16, RÉG 11]. The two main aspects of this activity are: knowledge organization processes (cataloging, indexing, social tagging, shelf arrangement, etc.) and knowledge organization systems (bibliographic description rules and formats, taxonomies, folksonomies, classifications, thesauri, semantic networks, ontologies, etc.) [HJØ 16, MAZ 18]. Knowledge organization systems (KOS) are schemas that organize concepts (represented by terms that come from common word usage) and the relationships between these concepts to reflect all human knowledge or knowledge related to a knowledge domain [GNO 16, p. 403]. By *knowledge domain*, we mean "a group with an ontological base that reveals an underlying teleology, a set of common hypotheses, epistemological consensus on methodological approaches, and social semantics" [SMI 12, p. 114]. Medical specialties, such as allergology or pediatrics, can be considered as such. The International Classification of Diseases (ICD) and the Medical Subject Headings (MeSH) are examples of medical knowledge organization systems (MKOS), which list and organize concepts within medical domains.

8.3. Methodology

The research for original, peer-reviewed articles and conference proceedings was conducted on March 25, 2020 and constitutes our corpus. We first manually searched the archive of the journal *Knowledge Organization* (1974–2019) and the conference proceedings *Advances in Knowledge Organization* (1990–2018), published by the International Society for Knowledge Organization (ISKO). Due to a modest number of relevant articles found in the archive (34 selected works), we extended the search by consulting the *Knowledge organization literature database*. The search engine in the database is available on the official ISKO website[1]. The database contains a bibliography related to knowledge organization from journals on information science, linguistics and computer science. To query the database, we used the terms "health" and "medicine". The queries provided 135 results. After removing duplicates and results not relevant to our approach, we retained 25 works. To complete and vary our sample, we went to LISTA (Library, Information Science & Technology Abstracts), a database of abstracts of publications in information science and technology, edited by EBSCO. The query terms used in the "title" field were composed of the adjectives "health" and "medical", combined with the names of different types of KOS [MAZ 18]: "(TI health OR TI medical) AND (TI vocabulary* OR TI gazetter+ OR TI dictionnar* OR TI glossar* OR TI terminolog* OR TI nomenclature* OR TI taxonom* OR TI classification+ OR TI subject headings OR TI thesaur* OR TI folksonom* OR TI conceptual scheme+ OR TI ontolog* OR TI semantic network+ OR TI knowledge graph+ OR TI knowledge organization)". We filtered the results by peer-reviewed scientific journal articles and obtained 122 publications. After selecting the relevant works and eliminating the duplicates, we retained 10 articles. To complete our sample, we also integrated two articles

1 ISKO, "KO literature", available at: https://www.isko.org/lit.html, June 26, 2020.

obtained during our literature search in a general search engine. The final corpus amounts to 71 publications (see Appendix).

We analyzed, evaluated and summarized the content of the documents in the corpus, using two reading grids, established as a result of the first exploratory readings: the first one is dedicated to theoretical approaches and evaluations, and the second one focuses on the design work of MKOSs. The progressive reading of the corpus according to a hermeneutic approach allowed us to organize the data, information and knowledge into categories and to connect them to each other in order to discover facts, regularities, relationships and phenomena. We identified the MKOSs discussed in the literature and collected information about them. The analysis of the MKOSs, with the help of the second grid, made it possible to illustrate certain statements with concrete examples. Consultation of some additional resources completed the report.

8.4. Presentation of the corpus of publications between 1960 and 2019

Among the 71 articles published in information and documentation sciences between 1968 and 2018, we found 11 publications published between 1960 and 1999 and 60 between 2000 and 2019. The significant increase in publications from 2000 onwards can be explained by the development of the medical Internet and the large-scale computerization of the health sector, which took place at that time [CAM 14, GON 07, pp. 14–55, THO 04] and attracted the attention of scholars in the discipline.

8.4.1. *Specialists working on the organization of knowledge in medicine*

Among the authors of our corpus, 70% are researchers, while 30% come from the professional sector. First, we identified information and documentation specialists. As far as scientists are concerned, they are affiliated with schools, faculties, colleges, as well as centers, departments and programs within universities specialized in Archivistics, Library & Information Science, Information Science, *Biblioteconomía y Documentación* (Librarianship and Documentation), Information and Documentation, Information Management, Communication & Arts, Communication Studies, Information and Media Studies and *Sciences de l'Information et de la Communication* (Information and Communication Sciences). On the professional side, they are often librarians, documentalists and catalogers. They work in national health libraries or in specialized libraries and resource centers, which are aimed at the general public, the academic community (higher level health institutions, universities) or hospital professionals. We also found articles published by employees working in private companies providing documentation services. Secondly, we identified co-authors who participated in projects conducted by information and documentation specialists. These are researchers attached to schools, centers and departments of informatics within universities, or affiliated with faculties of medicine and research institutes in epidemiology and public health. We also identified computer scientists and physicians who work in university hospitals and healthcare centers.

The majority of authors (58%) come from countries in the Americas: the United States leads, followed by Canada, Brazil, Mexico and Puerto Rico. The minority (27%) come from European countries, in particular from Western Europe (France, Spain, United Kingdom, Italy, Germany, Switzerland) and Northern Europe (Sweden, Finland, Norway, Denmark). They are followed by specialists from Asian countries (14%) (India, Japan, Singapore, Israel, China, South Korea) and Australia (1%). No work from an African country has been identified.

8.4.2. *The socio-documentary fields studied*

The research first focuses on techniques and MKOSs for organizing resources within institutions that most commonly hold documents: national and public libraries, general university libraries, health libraries and documentation centers [Texts 11, 13, 16, 37, 46, 65].

Following the computerization of library catalogs and the creation of medical bibliographies in the 1970s [NAT 19], and then the development of the Web from the second half of the 1990s [GON 07], specialists in the field have been interested in the evaluation and design of MKOSs and indexing methods in these environments [Texts 3, 4, 6, 10, 19, 20, 21, 22, 23, 25, 35, 36, 41, 42, 52, 53, 55, 62, 65, 66].

After the computerization of health systems in Western countries at the turn of the 1990s and 2000s [CAM 14, GON 07, pp. 14–55, THO 04], researchers have been focused on healthcare institutions. These are reflections on indexing practices and on the representation of documents recording activity (notably patient records) and information resources (scientific articles) as well as the design of ontologies for the development of knowledge bases, essential for the practices of health professionals [Texts 2, 5, 8, 15, 17, 31, 48, 49, 60].

The widespread availability of health information on the medical Internet at the turn of the 2000s and 2010s [ROM 08] has contributed to the extension of the field of investigation to the webosphere. Nowadays, scholars study methods of organizing medical knowledge and the info-communicative activities of users on websites and social media [Texts 1, 7, 9, 24, 18, 24, 28, 29, 30, 32, 34, 35, 43, 44, 47, 51, 56, 61, 69, 70].

Recently, information and documentation sciences have been involved in public health research and medical knowledge engineering, trying to extract knowledge from massively accessible information on the Internet (big data). These are studies that apply the techniques of literature-based knowledge discovery, sentiment analysis and information visualization (dataviz) in the processing of data from social media, the online press or bibliographic databases [Texts 33, 39, 59, 61].

8.5. Medical knowledge organization systems

Medical knowledge organization systems (MKOSs) are schemes consisting of concepts and relationships between concepts, representing knowledge about different domains of medicine and related fields. In MKOSs, concepts correspond to terms that are derived from specialized vocabulary (e.g. "seasonal allergic rhinitis") and/or lay terminology (e.g. "hay fever"). Furthermore, as evidenced by the analysis of the MKOSs in our corpus, diseases are often classified by reference to the human body system, for example, the cardiovascular, gastrointestinal, respiratory or immune system. Information and documentation sciences are mainly involved in the organization of knowledge in Western medicine. However, as complementary and alternative medicine is gaining popularity among the general public, it is also beginning to be considered. Examples are Ayurveda, Siddha medicine, traditional Chinese medicine, aromatherapy and herbal medicine [Texts 30, 31, 64, 67, 68].

8.5.1. *Typology and uses*

Taking into consideration the level of specialization of the topics covered by the MKOSs and the purpose of the devices for which they are implemented (general or specific audience), we can distinguish between general and specialized systems. General MKOSs aim to index all human knowledge and devote part of their schemas to concepts representing medical knowledge. They are used more in services and devices of a multidisciplinary nature, intended for a large, often non-specialized public. Specialized MKOSs emerged at a time when the rapid growth in the volume of medical knowledge, the constant evolution of medicine and the emergence of new specialties at the turn of the 20th century, required the development of specialized tools, sufficiently detailed to satisfy the informational needs of users of scientific and technical information [Texts 46, 50, 54]. Specialized systems are mainly dedicated to phenomena in medical fields but also include associated disciplines. They are mobilized in organizations, in systems and in documentation services specialized in medicine and health, aimed mainly at professionals of the sector, as well as the general public. Their typology has not changed since the 1960s [Text 46] – they are either very general specialized systems that include all fields of medical sciences, such as the National Library of Medicine (NLM) Classification and the Medical Subject Headings (MeSH), or they are MKOSs, dedicated to limited areas such as the Thesaurus of Psychological Index Terms and the Colon Cancer Treatment Ontology. Domain MKOSs often focus on medical specialties, for example: obstetrics and neonatology [Text 2], pharmacy [Text 37], neurosurgery [Text 49], nursing [Text 23], as well as on specific pathologies: brain tumors [Text 17], colon cancer [Text 33, 45], infectious and parasitic diseases [Text 52], AIDS [Text 27]. The designers also focused on important public health issues, such as health insurance [Text 11, 60], health and aging [Text 55] or nutrition [Text 71].

The analysis of our corpus highlights the use of different types of MKOSs, including: general classifications, medical classifications, medical nomenclatures, subject headings, medical thesauri, semantic networks, medical ontologies, medical glossaries, dictionaries and terminology directories, categorization schemes, medical taxonomies, folksonomies and consumer health vocabularies.

General classifications can be divided into enumerative classifications (Dewey Decimal Classification, Universal Decimal Classification, Library of

Congress Classification) and faceted classifications (Bliss Bibliographic Classification, Colon Classification). They have been developed since the 19th century to classify medical and health collections in public libraries, in general university libraries and also, though more rarely, in university health libraries which are part of the larger university library system [Texts 11, 16, 37, 65].

As noted in Texts 13, 16, 37 and 65, medical bibliographic classifications are used in health libraries, located in medical schools or hospitals (NLM Classification), and in health sections created in public libraries (Plantree Classification), for the classification of openly accessible materials. It is interesting to note that between the 1950s and 1970s, there was a wide range of medical classifications in health libraries (Cunningham Classification, Boston Classification, Barnard Classification, Army Medical Library Classification, etc.) [Text 46], whereas nowadays, the NLM Classification and general classifications significantly dominate the sector [Text 65]. Other medical classifications (International Classification of Diseases – ICD, Diagnostic and Statistical Manual of Mental Disorders – DSM, *Classification commune des actes médicaux* and nomenclatures (SNOMED-CT, RxNorm) are mobilized to index the resources contained in electronic patient records [Texts 2, 15, 48]. For example, the ICD is used for coding morbidity to generate epidemiological statistics on a global scale, while the *Classification commune des actes médicaux* (Common Classification of Medical Procedures) is used in France for medico-economic purposes, in particular for the coding of medical and technical procedures performed by health professionals. Furthermore, medical classifications and nomenclatures are considered terminological standards that ensure semantic interoperability between health data repositories. They therefore become components of the shared medical record, which allows patients to share their data with healthcare professionals[2] [ODE 16, Text 15].

Subject headings (Library of Congress Subject Headings, RAMEAU, *Répertoire de vedettes-matière de l'Université de Laval*), developed between the end of the 19th and the second half of the 20th century, are used by documentation professionals to describe and index the content of documents in the catalogs of public, general university and health libraries [Texts 10, 41, 65].

2 WHO, "Classifications", available at: https://www.who.int/classifications/icd/en/, 2020.

Medical thesauri (MeSH, Emtree, Thesaurus of Psychological Index Terms), semantic networks (UMLS) and, more recently, ontologies (OntoDip) are developed and used by information professionals and computer programs to describe and index the content of documents in online library catalogs, bibliographic databases (e.g. MEDLINE/PubMed) and strategic information systems [Texts 3, 4, 6, 10, 20, 23, 25, 32, 37, 51, 52, 53, 65, 66]. Moreover, ontologies, which allow the "conceptual modeling of a domain and the definition of a shared vocabulary", as well as the "simulation of human reasoning" [WAL 18, p. 163], act as knowledge bases and are used for semantic information retrieval and for knowledge extraction from electronic patient records and scientific literature [Texts 17, 31, 33, 39, 49]. As evidenced by work in computer science, ontologies are also being mobilized for medical imaging management and retrieval [MAR 11]. Information and documentation specialists also design glossaries, dictionaries (lexicons) and directories of medical terminologies [Texts 3 and 44]. These are schemes, often organized alphabetically, that list specialized or lay vocabularies used in a medical field, which can be used as terminology resources in language processing systems based on a linguistic approach or for the creation of ontologies.

Finally, some MKOSs (categorization schemes, taxonomies, consumer health vocabularies, ontologies) are used for the organization and retrieval of specialized and general public information on the Web. A categorization scheme is used on the National Center for Complementary and Integrative Health website[3]. The faceted Health Topics taxonomy, which organizes topics related to common diseases and conditions, mediate content on MedlinePlus [Text 51]. Folksonomies are tags, keywords and hashtags created by social media users through the process of social tagging [Texts 1, 18, 34, 35]. Consumer Health Vocabularies (CHVs) are schemas used to represent lay terminology, taking the structure of different MKOSs (from a terminology repository, classification, thesaurus or ontology), and are used, for example, in search engines for consumer services [Texts 9, 44, 56]. As highlighted by work in computer science and biostatistics, CHVs are also used in text mining tools when automatically processing corpora of data from social media [NZA 15].

3 National Institutes of Health, "National Center for Complementary and Integrative Health", available at: https://www.nccih.nih.gov/, 2020.

8.5.2. *The sources of the description*

MKOSs primarily aim to represent the content of documents by expressing their subject through a concept, represented by a term, sometimes accompanied by alphanumeric symbols (notations) or graphics (images). The subject may concern a discipline to which the document refers: *L9F:4* "Gynecology" (Colon Classification), or a phenomenon discussed in the document, for example a disease: *EA80* "Atopic Eczema" (ICD-11) or a treatment: *QW 900* "Desensitization, Immunologic" (NLM Classification). We can therefore consider them as *content models* of documents [ZAC 10, p. 148]. Moreover, with some specialized systems, it is possible to model *extra-documentary situations* [*ibid.*, p. 148]. In the MeSH, edited by the NLM, for example, we find the descriptor "Publication Characteristics" *[V]* which means the form of documents can be described, in particular the "Publication Formats" (e.g. "Journal Article", "Conference Proceedings", "Academic Dissertation"), "Study Characteristics" (e.g. "Clinical Study", "Meta-Analysis", "Multicenter Study" and "Publication Components" (e.g. "Comment", "Advertisement", "Published Erratum"). Classifying and sorting publications in medicine according to these criteria is important for the practice of evidence-based medicine, where the type of publication and study methodology determines the level of evidence in the work [HAS 13, HJØ 16]. Furthermore, in Siddha Ontology, modeled by Hemalata Iyer and K. S. Raghavan, we find concepts related to institutions, domain experts and types of collections and documents: reference books, journals, articles and manuscripts in the domain [Text 31]. Moreover, in MKOS, users can be grouped into sociological categories according to the type of information or collection that may satisfy their info-communication needs and practices. For example, the categorization scheme used on the website of the National Center for Complementary and Integrative Health contains the categories "Be an informed consumer" and "For Health Care Professionals" which address the specific types of users to whom the service is dedicated.

8.5.3. *The design of new MKOSs*

Our corpus includes 22 original articles on the design of MKOSs. The most frequently developed systems are ontologies (10 works) [Texts 2, 9, 17, 31, 33, 45, 49, 52, 60, 71] in the context of researchers' work, then thesauri (7 works) [Texts 6, 10, 23, 25, 27, 55, 67] and finally classifications (3 works) [Texts 11, 40, 50], generally developed and updated by documentation professionals. We also note the design of a conceptual framework of complementary and alternative medicine to build bridges between the tradition of Western medicine and the tradition of Eastern medicine [Text 68]. In addition, one work proposes the development of a repertoire of lay medical terminology [Text 44].

8.5.3.1. *The designers*

As the texts in our corpus attest, researchers in information and documentation sciences most often develop conceptual models [Texts 2, 9, 17, 31, 33, 49, 50, 60, 67, 71], as well as formalized systems [Texts 2 and 17]. The development of organizational systems, ready to be implemented in information systems (e.g. in an integrated library management system or in a shelf signage system of a library), is more the preserve of information professionals [Texts 6, 10, 13, 23, 25, 55].

Some MKOSs are developed outside the information and documentation community. The ICD, for example, is revised and maintained by the World Health Organization, in collaboration with medical experts in various fields. SNOMED-CT is developed by a non-profit organization, SNOMED International, which brings together specialists in information and communication technologies, information architecture and medicine. Moreover, as shown by the programs of scientific events around artificial intelligence and knowledge engineering, organized in France and internationally, notably the *Journées francophones d'Ingénierie des Connaissances*[4] (*Atelier IA et Santé*), the *Plateforme de l'Intelligence Artificielle*[5] (*Journée Santé & IA*) and the Congress on Medical and Health Informatics [OHN 19], laboratories, academic societies and consortia of informatics and artificial intelligence seem to be settling in the field of medical knowledge organization. By observing the different sources of funding for these events (governments, territorial communities, academic societies, universities, research organizations, scientific consortia, private companies and ICT giants), listed on the websites and mentioned in the works published in the proceedings, it seems that computer science and artificial intelligence have found a lucrative field of application, in which both private and public actors are interested, since the economic and societal

4 IC 2016, "Programme – 6 juin : IA et Santé 2016", available at: https://ic2016.sciencesconf.org/resource/page/id/27.html, 2016.

5 PFIA 2020, "Journée Santé et IA", available at: http://pfia2020.fr/journeesthematiques/sante-et-ia/, 2020.

stakes are very high. This funding allows them to conduct and present research on ontologies, terminologies and associated services. They rely on the exploitation of large corpora of texts and resort to automatic natural language processing (NLP) techniques, on the assumption that these corpora "provide the stable, consensual and shared elements of a domain" [AUS 12, p. 252]. Probably, thanks to the increasing amount of possible funding in this domain, we have assisted, over recent years, in increasing the interest from specialists in the field of information and documentation to develop medical ontologies. This trend is interesting as, although already conceptualized in computer science by Gruber in 1993 [GRU 93], ontologies are not even mentioned in the *Dictionnaire encyclopédique de l'information et de la documentation*, published in France in 1997 by Serge Cacaly [CAC 97]. As the publications in our corpus show, information and documentation specialists are now very much involved in the development and research of medical ontologies [Texts 2, 9, 17, 31, 33, 45, 49, 52, 60, 71]. In addition, Texts 2, 9, 52 and 60 confirm the funding of ontology projects led by information scientists, ministries of education, public health organizations, scientific development agencies and scientific associations.

8.5.3.2. *Design methods*

Developing MKOSs involves the identification of different elements relevant to a domain: concepts, terms, situations, objectives and issues, through analysis of different textual corpora. For this purpose, different types of documents are used: books, abstracts, scientific or general public articles, clinical documents, social media messages, websites of academic societies, query logs, descriptors of indexed documents, other MKOSs and survey data [Texts 2, 9, 10, 13, 17, 23, 25, 27, 31, 33, 40, 44, 45, 49, 50, 55, 56, 60, 67, 68]. Recently, authors have also used crowdsourcing involving non-experts [Text 71]. Furthermore, as pointed out in Texts 49 and 56, there are ethical issues that need to be addressed when designing MKOSs, especially when extracting terms from the contents of computerized patient records and social media, which include personal data. After the collection of relevant elements for a given domain, the concepts and terms expressing them are then normalized and organized by professionals, or by a machine, into semantic relation systems.

Information and documentation scholars' methods of designing MKOSs can be attributed to five different epistemological perspectives: pragmatism, historicism and hermeneutics, socioconstructivism, rationalism and empiricism [GNO 20, pp. 39–40, HJØ 03, p. 107, HUB 98, WEI 16].

8.5.3.2.1. The pragmatic method

Analysis of the objectives, values and application guidelines of a community, which guide the production and use of genres of discourse in a field (hospital reports, scientific articles, messages on social media, etc.), is the first method attached to pragmatism [GNO 20, p. 39]. It involves the study of the uses of health information, seen as social, communicational and cultural practices [Text 58]. Most often, this involves using the categories of information searches, identified through studies of users' information behavior [Text 57]. For example, Hemalata Iyer conducts interviews with practitioners and educators of Siddha medicine to identify the principles, professional and informational practices, communication patterns and knowledge structures developed, and then transposes this information into a domain ontology model [Text 31].

8.5.3.2.2. The historicist and hermeneutic method

The second method is part of the historicist or hermeneutic perspective. It involves studying the context and development of knowledge and the social and scientific division of labor, and categorizing concepts from a cultural point of view [GNO 20, p. 39, HJØ 03, p. 107]. The Universal Decimal Classification, the Bliss Bibliographic Classification and the NLM Classification are thus conceived and revised by librarians and documentalists in Europe and America who organize medical knowledge from the perspective of Western medicine. As a result, the fields of complementary and alternative medicine, which often belong to Eastern traditions, are poorly represented [Texts 30, 64]. It also seems that the MKOSs constructed in Eastern cultures are influenced by Western medicine. The Colon Classification, for example, which is revised in India, is not sufficiently developed to cover collections on Indian Ayurveda medicine [Text 30].

8.5.3.2.3. The socioconstructivist method

The third method, belonging to socioconstructivism, consists of the modeling of classification schemes that reflect knowledge influenced by social constructions, developed through mutual actions between individuals and their environment [HUB 98, WEI 16]. These constructs, which derive from individual experiences and social perspectives (formed, for example, by social ideologies or political interests), are said to stimulate the representation of a reality, which can sometimes be disconnected from scientific reasoning [HUB 98, pp. 195–196]. In the field of HIV/AIDS, for example, diseases and associated phenomena coexist together in such a reality, because the categorization, marginalization and stigmatization of populations affected by HIV (gay men, black people, drug users, etc.) or of their sexual behaviors (considered in some discourses as deviant or immoral) are influenced by social stimuli [*ibid.*, pp. 195–201]. According to a socioconstructivist approach, MKOSs must reflect the complexity of realities marked by the coexistence of social constructions alongside historical and cultural factors. To develop mediating vocabulary to support dialogue in the HIV/AIDS community, as an attempt to accommodate such complexity, Jeffrey F. Huber and Mary L. Gillaspy look at the different models of communication and knowledge production (both traditional and non-traditional), as well as the different vernaculars developed by different actors (clinicians, researchers, patients, etc.). Given the multidisciplinary, interdisciplinary and transdisciplinary nature of the field, Huber and Gillaspy base their vocabulary on the different domains: "[...] Epidemiology and Transmission; Education and Prevention; [...] Treatments, Therapies, and Medical Management of HIV Disease; Psychosocial and Religious Issues, Case Management; Legal, Ethical, Economic, and Political Aspects; [...] Fine Arts; Belles Lettres and Non-Fiction" to which they attribute relevant concepts [Text 27, p. 297]. They also use the common facets: "Age Ranges, Sexual Orientation, Gender, Stages of Infection, Ethnic Groups, Geographic Names, At-Risk Populations, Religious Faiths, Signs and Symptoms, and Special Populations", with a syntactic role that allows them to be associated with the main categories, and to complete them, if necessary [HUB 98, p. 203]. It is interesting to observe that, although the work is based on a socioconstructivist perspective, the designers finally resort to facet analysis, which makes it possible to see that their method is also part of a rationalist approach, another method of classification which is presented below.

8.5.3.2.4. The rationalist method

The fourth method is to create logical divisions that are considered universal [GNO 20, p. 39, HJØ 03, p. 107]. This method of organizing concepts is often based on the theory of facet analysis. It is considered by Birger Hjørland to be an expression of rationalism [HJØ 14], even if other authors, such as Joseph T. Tennis [TEN 08], also see components of pragmatism in it, because the fundamental categories used to organize knowledge (which we explain later) prove useful for the classification process, and classifications that rely on these categories save the reader time. Moreover, even if we can criticize the confusion between pragmatism and the practicality of this method (practicalism), it seems to us that it is primarily the facet analysis, based on the contents exchanged and used in a domain (scientific articles, books, patient files, etc.) for a concrete application, which allows us to see that it has pragmatist influences. The principles of this approach are based on a general theory of classification conceived by the Indian librarian Shiyali Ramamrita Ranganathan, and these principles were introduced in Europe in the 1950s by the Classification Research Group [VIC 60]. Based on a review of a sample of literature in a specialized field, concepts and terms are grouped into facets, which generally fulfill different semantic and syntactic roles. In faceted classifications, for example, concepts are assigned to facets of medicine. As Vanda Broughton points out, in class *H* "Physical anthropology, human biology, health sciences" in the second edition of the Bliss Bibliographic Classification, terms are assigned to the different categories:

– (Thing) human beings;

– (Kind) women, children, old people, etc.;

– (Part) head, legs, muscles, bones, heart, brain, lungs, etc.;

– (Process) respiration, digestion, reproduction, disease, etc.;

– (Operation) surgery, drug therapy, physiotherapy, etc.;

– (Agent) doctors, nurses, equipment, buildings, etc. [BRO 15, p. 308].

Thus, the terms in each category constitute a facet of a subject. In this way, within Bliss, all the terms of the category "Part", for example, become the facet of the class *HDP* "Anatomy" or *HPT* "PARTS, ORGANS, SYSTEMS OF THE BODY", while the terms of the category "Operation" become the facet of the class *HNP* "Treatment, therapy" [*ibid*]. In the same approach, K. S. Raghavan and C. Sajana propose facets of the domain, grouped into three major facets (which refer to Ranganathan's categories) to develop an ontology of neurosurgery dedicated to health professionals: "Personality (Patients, Body Organs, Health Care Personnel, Institutions, Hospital Managers), Matter-Property (Diseases, Symptoms, After Effects), Energy (Diagnosis, Treatment)" [Text 49, p. 210]. Since the ontology's mission is to ensure rapid access to clinical documents, Raghavan and

Sajana develop these categories and associate concepts with them based on the patient records produced and used by professionals in their professional practices. The mobilization of information content used by the actors in their daily practice to design an MKO, allows us to see that this approach does indeed have pragmatic aspirations.

8.5.3.2.5. The empirical method

The fifth method, which is empirical, is manifested by statistical generalizations, based on the "similarity" of informational objects that share a large number of properties in common [GNO 20, p. 39, HJØ 03, p. 107]. As evidenced in Texts 2, 9, 44, the empirical method often involves analysis of very large text corpora via NLP techniques to establish semantic relationships between ontology terms, or via machine learning (ML) techniques to establish links between terms and terminology definitions. An exemplary study of this approach is the work of Yunseon Choi, which focuses on building an ontology dedicated to healthcare consumers. She identifies a non-expert vocabulary via a latent semantic analysis of social tags, assigned by users of del.icio.us (a social bookmarking site) to index and share documents [Text 9]. It is interesting to observe that even if the neutrality of the statistical technique emanates from empiricism, the consideration of lay terminology in order for the ontology to best satisfy the needs of users, related to the search for relevant information resources, is in our view a pragmatist influence. On the other hand, the development of an ontology representing a non-expert view of the domain, based on a lay language established by an online community to index documents, also reflects socioconstructivist aspirations.

The authors rarely state the theoretical frameworks on which their work is based. One explanation lies in the recent nature of the epistemological foundation of MKOSs, developed with the formalization of the field of knowledge organization [GNO 20, HJØ 18].

8.5.3.3. *Test, validation and evaluation methods*

The knowledge organizations, proposed by information and documentation specialists, are tested, validated and evaluated in a qualitative and quantitative way. In the qualitative method, which is practiced most often, the aim is to evaluate the usefulness and efficiency of the MKOSs in representing the terms of the targeted domains or verifying their efficiency and their potential for integrating the information systems to which they are dedicated [Texts 2, 9, 17, 31, 49, 50]. The designers validate the designed MKOSs by conducting semi-structured interviews with individuals representing the categories of future or potential users: healthcare consumers [Text 9] or domain specialists [Text 2]. They are asked, for example, to indicate in a document the information they would like to have in a specific situation and the examiners see if these elements are correctly represented in the system being evaluated. MKOSs are also validated through text analysis, performed by experts, information professionals or by machines [Texts 2 and 50]. To validate the OntoNeo ontology, Mauricio Barcellos Almeida and Fernanda Farinelli use NLP techniques to automatically extract obstetrics and neonatology terminology from clinical records, which allows them to verify whether the most frequent terms, obtained from the extraction, are present in the ontology [Text 2]. In other works, designers submit queries to test the efficiency and potential functionality of information systems, based on MKOSs [Texts 17, 31, 49]. For example, the authors of the Siddha ontology perform semantic information retrieval in the knowledge base based on this ontology [Text 31]. The designers of the ontology of the neurosurgery domain, who rely on a decision support system based on the semantic indexing of patient records, ask this system specific queries, such as: "identify & retrieve records of patients: age = > 40 with astrocytoma" [Text 49, p. 214]. Quantitative evaluation aims to measure the number of documents retrieved via a query to the information system. At the end of this operation, the recall and precision rates are calculated [Text 60]. In addition, to evaluate the accuracy of the statements in the nutrition ontology, the recruited subjects evaluate the accuracy rate of the relations established by the designers [Text 71].

8.5.3.4. *New challenges for the design of MKOSs*

General and specialized classifications, in which knowledge is organized by discipline, are now confronted with the challenge of interdisciplinary knowledge. However, some interdisciplinary concepts from the medical sciences may appear in several classes within these classifications, since they are studied by several disciplines. In the NLM Classification, for example, the concept of "tumor" is found in different disciplines: pathology, ophthalmology, gynecology, nursing, etc. Such a classification method leads to the dispersion of resources, illustrated by the location of documents (on interdisciplinary topics) in several spaces of a library's classification scheme. As a result, there is a risk of dispersing material on different shelves of a library and, in this way, users may miss some of the resources that may be of interest to them. The somewhat dated (1996) but very timely study by Lynn Silipigni Connoway and Mary Ellen C. Sievert masterfully illustrates this problem: because of the interdisciplinary nature of the concept of "health insurance", resources on this topic, cataloged in bibliographic databases, are dispersed in 13 different classes of the NLM Classification and the Dewey Decimal Classification, as well as in 23 classes of the Library of Congress Classification [Text 11]. We see later that such inconsistency in classifications can create constraints for knowledge representation in Semantic Web standards.

Now, knowledge organization specialists recommend moving KOSs to online applications in order to optimize indexing and information retrieval in this environment [BIN 20]. Such a goal leads to representing the traditional organization of concepts in the syntax of Semantic Web technologies. While medical ontologies are natively created following these standards, MKOSs, whose origins date back to the 19th century and the beginnings of digital transformation, must undergo conversion to Networked Knowledge Organization Systems (NKOS). Following these recommendations, the Library of Congress and the NLM, respectively, published the Library of Congress Subject Headings and the MeSH in SKOS (Simple Knowledge Organization System) format, which is a standard and a set of specifications developed to support the use of MKOSs within the Semantic Web [Text 6, W3C 09]. However, in our opinion, basing general and specialized medical classifications on disciplines can become a considerable technical constraint for the conversion of these classifications to NKOS. Yet, for publishing classifications in SKOS format, the World Wide Web Consortium (W3C) recommends identifying MKOS concepts by URIs (Uniform Resource Identifiers), which allows concepts to be identified by a unique address for description and linking, using Resource Description Framework (RDF) or RDF Schema (RDFS) languages, based on the "subject–predicate–object" triplet model [W3C 09]. This specification seems difficult to achieve with disciplinary classifications where some interdisciplinary concepts may appear several times in a schema, and each time from a different perspective. A group of researchers from different countries working on the Integrative Levels Classification (ILC), a faceted interdisciplinary classification, proposes an alternative approach: instead of integrating interdisciplinary concepts into different disciplinary classes, they would represent them through phenomena, which are basic units of knowledge, independent of any discipline [GNO 16]. The latest work by this team on the conversion of ILC into SKOS format underlines that the phenomenon-based classification approach is relatively suitable for the representation of classifications in Semantic Web standards [BIN 20].

Other texts present reflections on the design and evaluation of MKOSs and bibliographic data exchange formats to improve knowledge discovery techniques based on scientific literature [Texts 33, 39, 59]. They are a continuation of the work initiated by Don R. Swanson on the establishment of new relationships between the knowledge contained in articles published in biomedical fields through treatments involving the use of semantic technologies. This idea can be related to the philosophy of the Web of data (linked data), which is a metadata infrastructure developed for the interoperability of repositories in order to "group the information in a useful way, as in a gigantic database where everything is written in a structured language", with the aim of developing different research services, consultation, exchange, annotation and indexing of contents [WAL 18, p. 159]. In order to describe the knowledge about the documentation resources and the relationships between these resources, it is necessary to use the languages of the Semantic Web that we have just mentioned (RDF, RDFS) and to build triplestores, which can be queried with the SPARQL language [*ibid.*, W3C 09].

Finally, the designers of MKOSs are currently required to organize information in a new type of medical device, notably in "healthcare interface organizations" [Text 5]. As French examples of such devices, we can mention the Personal Health Record, the Territorial Support Platform or telemedicine tools. The aim of their development is to associate the public and private sectors, including patients and their families, so that they become spaces for cooperation around shared knowledge. As Christian Bourret points out in Text 5, such systems require new thinking on the organization and representation of knowledge, adapted to the informational needs and skills as well as to the cognitive capacities of communities of scholars, experts, professionals and laypeople, in the spirit of building a collective intelligence based on the knowledge of all of their members.

8.6. New challenges for indexing and information retrieval

Part of our corpus of works focuses on the new challenges for indexing and retrieving medical and health information in the context of the social Web, the medical Internet and bibliographic databases.

The first category of publications deals with social indexing practices [Texts 1, 9, 18, 28, 32, 34, 35], "allowing non-specialist users to index documentation resources" on Web 2.0 [WAL 18, p. 161]. In the netnographic approach, specialists collect and analyze data with qualitative (observation, interview, facet analysis, comparison, equivalence, NLT) and quantitative (descriptive statistics, infometric and similarity measures) techniques. The results of these works highlight the subjective nature of social indexing, performed by different categories of users on online services. The study by Hemalata Iyer and Lucy Bungo, for example, shows that less than 1% of the tags concerning complementary and alternative medicine, posted by users of a social cataloging Web application, correspond to terms of the UMLS semantic network [Text 28]. The data collected by Margaret E. I. Kipp on CiteULike, a web-based bibliographic cataloging application, compared with those collected on PubMed, highlight that there are significant differences in the criteria for indexing biomedical literature between biomedical researchers and professional catalogers [Text 34]. Marit Kristine Ådland and Marianne Lykke emphasize that users and editors of a cancer website disagree on both the purpose of tags and what characterizes a good tag [Text 1].

Many MKOSs have been designed for information professionals to simplify access to information. However, the above-mentioned work points out that the medical terminology used by users and their indexing criteria may differ from the terminologies and criteria used by professionals. For example, Pertti Vakkari evaluates the extent to which the vocabulary used by users of digital question-and-answer services on the Web concerning health, nutrition and social services, is covered by general and specialized systems developed by information professionals. The results vary between 15% and 65% of terminological convergence, which shows a semantic divide between professional MKOSs and the vocabulary used by non-specialized users on the Web [Text 63]. These results make us think about the knowledge organization tools developed in a top-down approach, which can be incompatible with user behavior and practices, which can have negative consequences on the user's information search process and distance them from the information they need. Solutions to these terminologies and indexing discrepancies are most often sought in the construction of terminology resources and consumer health vocabularies [Texts 9, 44, 56] and in taking into account users' preferences for document indexing and information relevance criteria [Texts 1, 14, 29, 34, 35]. This kind of user knowledge can be acquired, for example, through domain analysis, which is a paradigm inspired by a sociological approach to knowledge organization, according to which indexing must be considered with a socio-cultural perspective, that is, linked to the social practices of individuals in a discourse community [HJØ 16]. Domain analysis thus brings together both hermeneutic and pragmatist approaches [GNO 20].

The second category of work is based on a cognitive approach to knowledge organization [GNO 20] which seeks to take into account the cognitive abilities of the user (located in a culture or community) when searching for information [Texts 24, 29, 47, 51, 56]. In this approach, Dagobert Soergel, Tony Tse and Laura Slaughter are inspired by the work on human cognition carried out in biomedical informatics, which emphasizes that the understanding of pathologies by patients and physicians can be summarized in two distinct models: the *illness model*, manifested in patients, and the *disease model*, characterizing the perception of physicians [Text 56]. Soergel's team bases their work on this model and develops an interpretive layer, which aims at mediating health information between the perspective of specialists and the perspective of non-experts. The model focuses on the design of a set of tools: an educational ontology (with non-expert definitions and terminologies), a validated biomedical knowledge base, a false-belief explanation base and a visual display and classification navigation interface. All of these tools form an "interpretative layer", allowing the non-expert user to find the document related to their information needs and to understand its content. In addition, to improve access to information, some designers seek to represent knowledge in a design appropriate to the cognitive abilities associated with the information culture of particular groups. For example, the work of Lisa M. Given *et al.* and Hemalata Iyer and Mary Guadrón attempts to identify strategies for organizing and visualizing knowledge on health websites for older adults [Texts 24 and 29]. Kyong Eun Oh, Soohyung Joo and Eun-Ja Jeong collect facets that are preferred, easy to use and useful, in order for the general public to access online health information [Text 47].

The third category of studies deals with the phenomenon of crowdsourcing [Text 66]. The study by James W. Woods *et al.* provides instructive feedback on this practice, using the example of the indexing of images of skin lesions from the NLM collection by dermatologists [Text 66]. Nine experts spent an average of 30 minutes describing the images via a form made available to them on the Web. Comparison of the vocabulary used for indexing, with the UMLS terms, shows that 300 terms from this

MKOS out of the 346 unique terms submitted by the practitioners have an exact equivalence [Text 66]. This example highlights the great potential of participatory production, which could be used for indexing and annotating a large corpora of clinical activity documents or health information resources (e.g. heritage collections). Crowdsourcing could also be an interesting alternative to automatic indexing and annotation techniques.

The fourth category of work shows that vocabulary control and rigid structuring of terminology in MKOSs built in a top-down approach, e.g. medical thesauri, are still useful for scientific information retrieval. Ying-Hsang Liu and Nina Wacholder point out that MeSH terms are indispensable for measuring the accuracy rate for domain experts, even for graduate students in biomedical fields who have not undergone specific training in information retrieval [Text 36]. In addition, Xiangming Mu, Kun Lu and Hohyon Ryu, who conducted a study of undergraduate and graduate students with no experience in information retrieval, show that a system that makes it possible to search by MeSH term and navigating its schema is more effective in information retrieval than a system that provides a simple search interface [Text 43]. Finally, as evidenced by numerous medical theses and articles, MeSH is currently widely used by both researcher-practitioners and medical students to query PubMed for systematic literature reviews.

8.7. Social, cultural, political and economic issues

In addition to the problems related to indexing and information retrieval, the organization of medical knowledge also raises social, cultural, political and economic issues.

As Texts 12 and 26 in our corpus attest, adding or removing concepts in an MKOS is based on a socio-political evolution of knowledge and raises sensitive questions and even political tensions, prompted by normative issues. In 1968, the American Psychiatric Association added homosexuality to the Diagnostic and Statistical Manual of Mental Disorders (a classification of mental illness). This appearance provoked protests from homosexual activists. After a long debate, in 1973, the term was removed from the classification [Texts 12 and 26]. Moreover, in the ninth version of the International Classification of Diseases (ICD-9), used until the end of the 1980s, the concept of homosexuality (*302.0* "Homosexuality") appeared in the scheme, as the first subdivision of class *302* "Sexual deviations and disorders", and belonged to chapter *300-316* "NEUROTIC DISORDERS, PERSONALITY DISORDERS, AND OTHER NONPSYCHOTIC MENTAL DISORDERS". The concept was accompanied by a usage note for health professionals: "Code homosexuality here, whether or not it is considered a mental disorder". Even if the classification never explicitly stated that homosexuality should be considered a disease, the analysis of generic concepts, allowing the context in the classifications to be revealed [ZAC 10], led to interpretations of the media discourse as discriminatory and erratic [VAN 94]. In the 10th version of the classification, published in 1990 to replace ICD-9, the term no longer appears as a reason for sexual contact [DRI 94]. The above examples clearly illustrate the normative stakes of MKOSs, which reflect the power of certain actors to categorize what is normal and what is pathological [WAL 18, p. 158]. The categorization of individuals has a direct link to their representation in a society. So much so that it is understandable that it becomes a tool of revolt for communities that do not identify with certain categorizations proposed by normative bodies.

Moreover, the cultural warrant principle, which calls for the design of MKOSs that are appropriate and useful to the individuals of a culture, may lead to the interests, values, rights and beliefs of others being overlooked, depriving them of the right to access information [BEG 02]. As we have already pointed out in the section on MKOS design methods, the fields of complementary and alternative medicine are poorly represented in the classifications reviewed by Western information professionals [Texts 30 and 64]. Information resources on Eastern medicine topics may therefore be less visible in the collections of a library located in Europe. As a result, followers of Siddha medicine who visit this library, for example, are less likely to find the information they need.

Finally, as emphasized by work in the information and communication sciences and in medicine, the design of reference medical classifications also raises economic issues, linked to political decisions taken by governmental bodies and intergovernmental organizations responsible for their revision [TAN 16, WAL 18, p. 158]. The inclusion or exclusion of certain diseases in classifications such as the ICD or the *Classification commune des actes médicaux* (French Common Classification of Medical Acts), for example, has repercussions on the marketing of new drugs by the pharmaceutical industry and on the reimbursement of medical acts by national health insurance funds, mutual health insurance companies and insurance companies. As an illustration of such power, we can mention a case reported by specialists in allergology, a recent specialty that is sometimes incorrectly represented in medical classifications. The absence of a specific class for anaphylaxis in ICD-10 due to the confusion of this disease with anaphylactic shock in class *T78.2* “Anaphylactic shock, unspecified” (the terms “anaphylaxis” and “anaphylactic shock” do not represent the same concept), and the downgrading of the concept of anaphylaxis to class *T78* “Adverse reactions, not elsewhere classified”, has led to the underreporting of anaphylaxis mortality by health professionals in Brazil when coding death certificates. The lack of epidemiological data in the Brazilian national database has significantly affected decision-making by the government and pharmaceutical companies and, as a consequence, the non-marketing of self-injectable epinephrine for the treatment of anaphylaxis [TAN 16].

The above examples make it clear that MKOSs represent and impose a certain worldview at a given moment and in a particular context [DUB 20, p. 12]. They can also direct knowledge, give visibility to a pathology, become instruments of power, revolt and access to information, as well as enable or prevent the circulation of goods in the health economy.

8.8. Discussion

Our analysis leads us to the observation that the theories and techniques for organizing medical knowledge, developed by information and documentation specialists, have been profoundly affected by the digital transformations of the health information field. On the technical side, the constant growth in the volume of knowledge in this field has created, for example, the need for specialized MKOSs. The paradigm of evidence-based medicine has influenced the criteria for indexing medical literature. The development of Web 2.0 has contributed to the emergence of the practice of social indexing and has created the need to adapt knowledge organization tools to this environment. In addition, the Semantic Web and the idea of Linked Data have guided the representation of MKOS in new computer standards and languages. Finally, massive data have attracted the attention of researchers who try to extract knowledge from them. In parallel to these technical aspects, we observe the development of theoretical reflections. Epistemological and methodological contributions as well as ethnographic and netnographic studies are the manifestation of this new approach, which seeks to put the user at the heart of the reflection. This appears to be a consequence of the participative and social character of Web 2.0, which has facilitated the obtaining of behavioral data, the analysis of which has led to an awareness of users' needs. First, the importance is given to the non-specialized user. Traditional medical terminologies and indexing techniques are being challenged. They are confronted with the vocabulary, indexing and navigation preferences, as well as the cognitive abilities of the users. Then, other works focus on the specialized user within their work context. This is illustrated by MKOS design methods that take into account different vernacular vocabularies built within medical domains, as well as the needs, behaviors and info-communication practices of health professionals. Finally, specialists are interested in the different issues of knowledge organization, a concern that is specific to the HSS.

Despite the creation of a corpus from only three data sources, the technical and theoretical trends are in line with the global changes affecting the field of knowledge organization [SOS 19, pp. 31–43, WAL 18, pp. 155–164]. The large-scale design of medical ontologies for online applications and for knowledge organization, management and retrieval processes is a leading example. Designers are evolving traditional systems to conform to Semantic Web standards. Participatory production is also spreading, little by little, for the construction of MKOSs and for the indexing of documentation resources on medical topics. Moreover, TALN, clustering and information visualization techniques are increasingly practiced, which may reflect a rapprochement of the fields of knowledge engineering and artificial intelligence, given the interdisciplinary nature of information sciences which oscillate between HSS and computer science, as well as, quite simply, a step towards the adaptation of knowledge organization techniques to new technological trends and requirements.

Moreover, statistical classification techniques, such as ML, are already used to develop predictive models (classification algorithms), for example, for cancer diagnosis, without recourse to classical knowledge organization tools [SID 19]. As a consequence, the relevance of MKOS description, classification and document indexing seems to be questioned with respect to ML techniques such as deep learning, artificial neural networks, etc. [SID 19]. On the other hand, Andrew MacFarlane, Sondess Missaoui and Sylwia Frankowska-Takhari point out that knowledge organization tools should not be neglected in this task [MAC 20]. Using the example of multimedia documents, the authors highlight that the implementation of projects involving supervised learning techniques first requires knowledge about the informational needs and practices in the domains of investigation (which can be mediated by KOSs) in order to identify the elements relevant to the parts of the document contents to be "learned" by the machines. Second, the algorithms used in this kind of work require training documents with metadata describing their "general" and "contextual" features (e.g. objects, styles, genres, emotions, locations), which are difficult for the algorithms to detect [MAC 20]. According to the model developed by MacFarlane *et al.*, training corpora for ML can involve documents and associated subject metadata (controlled by KOSs), for example, from image banks. They can then be merged with image corpora, whose features can be extracted using ML (machine learning) techniques, to create the indexes needed for automatic classification. The relevance of this model for the classification of information in the medical domain remains to be evaluated.

The positioning of information and documentation sciences with regard to the organization of medical knowledge can be explained by taking into consideration general and specialized MKOs and by analyzing different facets of medical fields. This approach allows for a holistic view of the objects of study. As we have just seen, the dimensions of knowledge that are often investigated are: reality, ontology, epistemology, documents, collections, information needs (connected to informational behaviors and info-communicational practices) and cognitive capacities of individuals. These "steps between the world and classification", which can be identified in the contents of MKOSs, are highlighted by Claudio Gnoli from an initial scheme developed by Brian Vickery [GNO 20]. They can be applied in medical domains as follows:

– α reality (the undifferentiated totality upstream of human understanding);

– β phenomenon, for example allergic hypersensitivity;

– γ perspective, for example an integrated approach;

– δ document, for example an article in a scientific journal that discusses it;

– ε collection, for example a digital library that contains the item;

– ζ information needs, for example a search by medical specialty thesis;

– η people, for example an intern in a teaching hospital;

– θ cognition, for example their natural abilities to navigate lists.

The works inspired by a pragmatic approach, analyzed in this study, also underline that some MKOS designers now feel the need to free themselves from a technocentric approach, which gives pride of place to the tools implemented in the health sector [GRO 07], and to build digital devices and services in connection with the users' practices. The study of medical domains in a holistic approach, specific to information and documentation sciences, thus seems very favorable. This approach can lead to the proposal of MKOS models for the design of mediation devices, which are then translated into concrete realizations by engineers. As researchers in information and communication sciences and professionals in information–documentation, we encourage specialists in the discipline to set up multidisciplinary projects in digital humanities, in partnership with the health sector and in cooperation with specialists in computer science.

8.9. Conclusion

The work in information and documentation sciences on the organization of medical knowledge is divided into two axes. The first one invites us to conceive and evaluate the MKOS. It is characterized by analyses of textual corpora, the use of documentary techniques, semantic technologies and through studies focused on users, their info-communication activities and their cognitive capacities. The second axis includes reflections specific to a science of mentefacts[6], notably on the theory and epistemology of knowledge organization and on the issues related to classification. In the first axis, the aim is to develop MKOSs for the design of systems allowing users to manage, access and consult information. In the second axis, medical discourse occupies a predominant place. Specialists analyze the discourse carried by information resources and MKOSs and emphasize their role as mediators of political, scientific, social, cultural and economic phenomena.

This study highlights that knowledge organization opens up a broad spectrum of applications in the field of medicine and health. It also helps explain why MKOSs, despite their artificial and often controlled nature, are able to draw society's attention to the sensitive nature of the health of the population and are powerful political instruments.

6 By the concept of "mentefacts" we mean the intellectual products of human creativity (culture): theories, MKOSs, books, critical thinking, etc. [GNO 19].

8.10. References

[AUS 12] AUSSENAC-GILLES N., CHARLET J., REYNAUD C., "Les enjeux de l'Ingénierie des Connaissances", in SEDES F., OGIER J.-M., MARQUIS P. (eds), *Information-Interaction-Intelligence : le point sur le I3*, Editions Cépaduès, Toulouse, 2012.

[BEG 02] BEGHTOL C., "A proposed ethical warrant for global knowledge representation and organization systems", *Journal of Documentation*, vol. 58, no. 5, pp. 507–532, 2002.

[BIN 20] BINDING C., GNOLI C., MERLI D. *et al.*, "Integrative levels classification as a networked KOS: A SKOS representation of ILC2", *Advances in Knowledge Organization, Vol. 17 (2020) – Knowledge Organization at the Interface – Proceedings of the Seventeenth International ISKO Conference 6–8 July 2020 Aalborg, Denmark*, Ergon, Baden-Baden, 2020.

[BRO 15] BROUGHTON V., *Essential Classification*, 2nd edition, Facet Publishing, London, 2015.

[CAC 97] CACALY S. (ed.), *Dictionnaire encyclopédique de l'information et de la documentation*, Editions Nathan, Paris, 1997.

[CAM 14] CAMPION-AWWAD O., HAYTON A., SMITH L. *et al.*, *The National Programme for IT in the NHS – A Case History*, University of Cambridge, Cambridge, 2014.

[COU 11] COUZINET V., "Les dispositifs : question documentaire", in GARDIÈS C. (ed.), *Approche de l'information documentation : concepts fondateurs*, Cépaduès, Toulouse, 2011.

[DUB 00] DUBRIDGE G., *Lost Books of Medieval China*, The British Library, London, 2000.

[GNO 16] GNOLI C., "Classifying phenomena, Part 1: Dimensions", *Knowledge Organization*, vol. 43, no. 6, pp. 403–415, 2016.

[GNO 19] GNOLI C., "Levels of information and library and information science as a science of mentefacts", *Information Research – Proceedings of CoLIS, the Tenth International Conference on Conceptions of Library and Information Science, Ljubljana, Slovenia, June 16–19, 2019*, vol. 24, no. 4, 2019.

[GNO 20] GNOLI C., *Introduction to Knowledge Organization*, Facet, London, 2020.

[GON 07] GONOD BOISSIN F., L'usage de l'information numérique en médecine générale : étude exploratoire en Rhône-Alpes, PhD thesis, Université Claude Bernard Lyon 1, 2007.

[GRO 07] GROSJEAN S., BONNEVILLE L., "Logiques d'implantation des TIC dans le secteur de la santé", *Revue française de gestion*, vol. 3, no. 172, pp. 145–157, 2007.

[GRU 93] GRUBER T.R., "Towards principles for the design of ontologies used for knowledge sharing", *International Journal Human-Computer Studies*, vol. 43, pp. 907–928, 1993.

[HAS 13] HAS – HAUTE AUTORITÉ DE SANTÉ, Niveau de preuve et gradations des recommandations de bonne pratique – Etat des lieux, available at: https://www.has-sante.fr/jcms/c_1600564/fr/niveau-de-preuve-et-gradation-des-recommandations-de-bonne-pratique-etat-des-lieux/, 14 June 2013.

[HJØ 03] HJØRLAND B., "Fundamentals for knowledge organization", *Knowledge Organization*, vol. 30, no. 2, pp. 87–111, 2003.

[HJØ 14] HJØRLAND B., "Is facet analysis based on rationalism? A discussion of Satija (1992), Tennis (2008), Herre (2013), Mazzocchi (2013b), and Dousa & Ibekwe-SanJuan (2014)", *Knowledge Organization*, vol. 41, no. 5, pp. 369–376, 2014.

[HJØ 16] HJØRLAND B., "Knowledge organization", *Knowledge Organization*, vol. 43, no. 6, pp. 475–484, 2016.

[HJØ 18] HJØRLAND B., "Library and information science (LIS). Part 2", *Knowledge Organization*, vol. 45, no. 4, pp. 319–338, 2018.

[HUB 98] HUBER J.T., GILLASPY M.L., "Social constructs and disease: Implications for a controlled vocabulary for HIV/AIDS", *Library Trends*, vol. 47, no. 2, pp. 190–208, 1998.

[MAC 20] MACFARLANE A., MISSAOUI S., FRANKOWSKA-TAKHARI S., "On machine learning and knowledge organization in multimedia information retrieval", *Knowledge Organization*, vol. 47, no. 1, pp. 45–55, 2020.

[MAR 11] MARAGOUDAKIS M., MAGLOGIANNIS I., "A medical ontology for intelligent web-based skin lesions image retrieval", *Health Informatics Journal*, vol. 17, no. 2, pp. 140–157, 2011.

[MAZ 18] MAZZOCCHI F., "Knowledge organization system (KOS)", *Knowledge Organization*, vol. 45, no. 1, pp. 54–78, 2018.

[MET 06] METZGER J.-P., "L'information-documentation", in OLIVESI S. (ed.), *Sciences de l'information et de la communication – Objets, savoirs, discipline*, Presses universitaires de Grenoble, Grenoble, 2006.

[NAT 19] NATIONAL LIBRARY OF MEDICINE, A brief history of NLM, available at: https://www.nlm.nih.gov/about/briefhistory.html/, 20 March 2019.

[NZA 15] NZALI M.D.T., BRINGAY S., LAVERGNE C., *et al.*, "Construction d'un vocabulaire patient/médecin dédié au cancer du sein à partir des médias sociaux", *26èmes journées francophones d'ingénierie des connaissances, Rennes, France, 1–3 July 2015*, available at: https://hal.inria.fr/IC-2015/, 2015.

[ODE 16] ODEH S., "Du dossier médical personnel au dossier partagé : vers un dispositif de médiation documentaire", *Les Cahiers du numérique*, vol. 12, no. 1, pp. 31–50, 2016.

[OHN 19] OHNO-MACHADO L., SÉROUSSI B. (eds), *MEDINFO 2019: Health and Wellbeing e-networks for All – Proceedings of the 17th World Congress on Medical and Health Informatics*, IOS Press BV, Amsterdam, 2019.

[RÉG 11] RÉGIMBEAU G., "3-Médiation", in GARDIÈS C. (ed.), *Approche de l'information documentation – concepts fondateurs*, Cépaduès, Toulouse, 2011.

[ROM 08] ROMEYER H., "TIC et santé : entre information médicale et information de santé", *Tic en santé*, vol. 2, no. 1, available at: https://journals.openedition.org/ticetsociete/365, 2008.

[SID 19] SIDEY-GIBBONS J., SIDEY-GIBBONS C., "Machine learning in medicine: A practical introduction", *BMC Medical Research Methodology*, vol. 19, available at: https://bmcmedresmethodol.biomedcentral.com/articles/10.1186/s12874-019-0681-4, 2019.

[SMI 12] SMIRAGLIA R., "Epistemology of domain analysis", in SMIRAGLIA R., LEE H. (eds), *Cultural Frames of Knowledge*, Ergon, Würzburg, 2012.

[SOS 19] SOSIŃSKA-KALATA B., "Organizacja wiedzy w cyfrowym świecie: Wpływ rewolucji cyfrowej na ewolucję metod i narzędzi organizacji wiedzy", *Zagadnienia Informacji Naukowej*, vol. 57, no. 1, pp. 26–46, 2019.

[TAN 16] TANNO L.K., CALDERON M.A., DEMOLY P., "New allergic and hypersensitivity conditions section in the international classification of diseases-11", *Allergy Asthma & Immunology Research*, vol. 8, no. 4, pp. 383–388, 2016.

[TEN 08] TENNIS J.T., "Epistemology, theory, and methodology in knowledge organization: Toward a classification, metatheory and research framework", *Knowledge Organization*, vol. 35, nos 2–3, pp. 102–112, 2008.

[THO 04] THOMPSON T.G., BRAILER D.J., *The Decade of Health Information Technology: Delivering Consumer-centric and Information-rich Health Care – Framework for Strategic Action*, U.S. Department of Health and Human Services, Office for the National Coordinator for Health Information Technology, Washington, D.C., 2004.

[W3C 09] W3C, SKOS: Simple knowledge organization system reference, available at: http://www.w3.org/TR/skos-reference/, 2009.

[WAL 18] WALTER J., DOUYÈRE D., BOUILLON J.-L. *et al.*, "Dynamiques des recherches en sciences de l'information et de la communication", *Conférence permanente des directeurs·trices des unités de recherche en sciences de l'information et de la communication (CPDirSIC)*, available at: https://pastel.archives-ouvertes.fr/CREM-UL/hal-01885229v2, 2018.

[WEI 16] WEISS L.C., BRÄSCHER M., VIANNA W.B., "Pragmatism, constructivism and knowledge organization", in CHAVES GUIMARÃES J.A., OLIVIERA MILANI S., DODEBEI V. (eds), *Advances in Knowledge Organization, Vol. 15 (2016) – Knowledge Organization for a Sustainable World: Challenges and Perspectives for Cultural, Scientific, and Technological Sharing in a Connected Society – Proceedings of the Fourteenth International ISKO Conference 27–29 September 2016 Rio de Janeiro, Brazil*, Ergon, Würzburg, 2016.

[VAN 94] VAN DRIMMELEN-KRABBE J.J., USTUN T.B., THOMPSON D.H. *et al.*, "Homosexuality in the international classification of diseases: A clarification", *JAMA*, vol. 272, no. 21, p. 1660, 1994.

[VIC 60] VICKERY B., *Faceted Classification: A Guide to the Construction and Use of Special Schemes*, ASLIB, London, 1960.

[ZAC 10] ZACKLAD M., "Evaluation des systèmes d'organisation des connaissances", *Les cahiers du numérique*, vol. 6, no. 3, pp. 133–166, 2010.

8.11. Appendix 1: the corpus of journal articles and conference proceedings

Text 1 – ÅDLAND M.K., LYKKE M., "Tags on healthcare information websites", in RIBEIRO F., CERVEIRA M.E. (eds), *Advances in Knowledge Organization, Vol. 16 (2018) – Challenges and Opportunities for Knowledge Organization in the Digital Age – Proceedings of the Fifteenth International ISKO Conference 9-11 July 2018 Porto, Portugal*, Ergon, Baden-Baden, 2018.

Text 2 – ALMEIDA BARCELLOS M., FARINELLI F., "Ontologies for the representation of electronic medical records: The obstetric and neonatal ontology", *Journal of the American Society for Information Science and Technology*, vol. 68, no. 11, pp. 2529-2542, 2017.

Text 3 – DE ANDRADE J., LOPEZ GINES DE LARA M., "Interoperability and mapping between knowledge organization systems: Metathesaurus – Unified Medical Language System of the National Library of Medicine", *Knowledge Organization*, vol. 43, no. 2, pp. 107-112, 2016.

Text 4 – BOEHR D., BUSHMAN B., "Preparing for the future: National Library of Medicine's® project to add MeSH® RDF URIs to its bibliographic and authority records", *Cataloging & Classification Quarterly*, vol. 56, nos 2-3, pp. 262-272, 2018.

Text 5 – BOURRET C., "New ways to produce shared knowledge to improve cooperation in overcoming societal challenges in healthcare: The lever of innovative interface organizations in France", in RIBEIRO F., CERVEIRA M.E. (eds), *Advances in Knowledge Organization, Vol. 16 (2018) – Challenges and Opportunities for Knowledge Organization in the Digital Age – Proceedings of the Fifteenth International ISKO Conference 9-11 July 2018 Porto, Portugal*, Ergon, Baden-Baden, 2018.

Text 6 – BUSHMAN B., ANDERSON D., FU G., "Transforming the medical subject headings into linked data: Creating the authorized version of MeSH in RDF", *Journal of Library Metadata*, vol. 15, nos 3-4, pp. 157-176, 2015.

Text 7 – CAMPBELL D.G., "Farradane's relational indexing and its relationship to hyperlinking in Alzheimer's information", in NEELAMEGHAN A., RAGHAVAN K.S. (eds), *Advances in Knowledge Organization, Vol. 12 (2012) – Categories, Contexts, and Relations in Knowledge Organization – Proceedings of the Twelfth International ISKO Conference 6-9 August 2012 Mysore, India*, Ergon-Verlag, Würzburg, 2012.

Text 8 – CAMPBELL D.G., "The human life as warrant: A facet analysis of protocols for dealing with responsive behaviours in dementia patients", in BABIK W. (ed.), *Advances in Knowledge Organization, vol. 14 (2014) – Knowledge Organization in the 21st Century: Between Historical Patterns and Future Prospects – Proceedings of the Thirteenth International ISKO Conference 19-22 May 2013 Kraków, Poland*, Ergon-Verlag, Würzburg, 2014.

Text 9 – CHOI Y., "Supporting better treatments for meeting health consumers' needs: Extracting semantics in social data for representing a consumer health ontology", *Information Research*, vol. 21, no. 4, 2016.

Text 10 – COGNETTI G., POPPI G., "The problems of indexing in Italian biomedical libraries", in NEGRINI G., FARNESI T., BENEDIKTSSON D. (eds), *Advances in Knowledge Organization, vol. 3 (1991) – Documentary Languages and Databases – Papers from the Rome Conference, December 3-4, 1990*, INDEKS Verlag, Frankfurt/Main, 1991.

Text 11 – CONNOWAY L.S., SIEVERT M.C., "Comparison of three classification systems for information on health insurance", *Cataloging & Classification Quarterly*, vol. 23, no. 2, pp. 89-104, 1996.

Text 12 – COOPER R., "Diagnostic and statistical manual of mental disorders (DSM)", *Knowledge Organization*, vol. 44, no. 8, pp. 668-676, 2017.

Text 13 – COSGROVE T.L., "Planetree health information services: Public access to the health information people want", *Bulletin of the Medical Library Association*, vol. 82, no. 1, pp. 57-63, 1994.

Text 14 – CRYSTAL A., GREENBERG J., "Relevance criteria identified by health information users during Web searches", *Journal of the American Society for Information Science and Technology*, vol. 57, no. 10, pp. 1368-1382, 2006.

Text 15 – CUENCA G.M., SALVADOR OLIVÁN J.A., "Representación del conocimiento en historia clínica electrónica interoperable: el caso de la Historia Clínica Digital del Sistema Nacional de Salud de España", *Scire: Representación y organización del conocimiento*, vol. 23, no. 1, pp. 25-38, 2017.

Text 16 – DALI K., DILEVKO J., "Towards improved collections in medical humanities: Fiction in academic health libraries", *The Journal of Academic Librarianship*, vol. 32, no. 3, pp. 259-273, 2006.

Text 17 – DAS S., ROY S., "Faceted ontological model for brain tumour study", *Knowledge Organization*, vol. 43, no. 1, pp. 3-12, 2016.

Text 18 – DEFRANCIS SUN B., "Folksonomy and health information access: How can social bookmarking assist seekers of online medical information?", *Journal of Hospital Librarianship*, vol. 8, no. 1, pp. 119-126, 2008.

Text 19 – DOUYÈRE M., SOUALMIA L.F., NÉVÉOL A., ROGOZAN A., DAHAMNA B., LEROY J.-P., THIRION B., DARMONI S.J., "Enhancing the MeSH thesaurus to retrieve French online health resources in a quality-controlled gateway", *Health Information and Libraries Journal*, vol. 21, no. 4, pp. 253-261, 2004.

Text 20 – EFTHIMIADIS E.N., AFIFI M., "Population groups: Indexing, coverage, and retrieval effectiveness of ethnically related health care issues in health sciences databases", *Bulletin of the Medical Library Association*, vol. 84, no. 3, pp. 386-396, 1996.

Text 21 – FERRAIOLI L., "An exploratory study of metadata creation in a health care agency", *Cataloging & Classification Quarterly*, vol. 40, nos 3-4, pp. 75-102, 2005.

Text 22 – FERNÁNDEZ L.M.M., IZQUIERDO ALONSO M., MAURANDI LÓPEZ A., VALLÉS VALENZUELA J., "Consistency between indexers in the LILAC database (Latin American and Caribbean Health Science Literature)", *Information Research*, vol. 18, no. 4, 2013.

Text 23 – FLOR P., JAKOBSSON A., MOGSET I., TAYLOR S., EASEN S.E., "A controlled vocabulary for nursing and allied health in Norway", *Health Information and Libraries Journal*, vol. 18, no. 1, pp. 10-19, 2001.

Text 24 – GIVEN L.M., RUECKER S., SIMPSON H., SADLER E.B., RUSKIN A., "Inclusive interface design for seniors: Image-browsing for a health information context", *Journal of the American Society for Information Science and Technology*, vol. 58, no. 11, pp. 1610-1617, 2007.

Text 25 – HAMADA M., SAKUMA S., MISAWA K., "Thesaurus for medical and health related terms and its relation to medical subject headings", *Journal of Information Processing and Management*, vol. 51, no. 9, pp. 642-652, 2008.

Text 26 – HUBER J.T., GILLASPY M.L., "An examination of the discourse of homosexuality as reflected in medical vocabularies, classificatory structures, and information resources", in BEGHTOL C., HOWARTH L.C., WILLIAMSON N.J. (eds), *Advances in Knowledge Organization, vol. 7 (2000) – Dynamism and Stability in Knowledge Organization – Proceedings of the Sixth International ISKO Conference 10-13 July 2000 Toronto, Canada*, ERGON Verlag, Würzburg, 2000.

Text 27 – HUBER J.T., GILLASPY M.L., "Controlled vocabulary for HIV/AIDS: An evolving nosological: record of a diseased body of knowledge", in GREEN R. (ed.), *Advances in Knowledge Organization, Vol. 5 (1996) – Knowledge Organization and Change – Proceedings of the Fourth International ISKO Conference 15-18 July 1996 Washington, DC*, INDEKS Verlag, Frankfurt/Main, 1996.

Text 28 – IYER H., BUNGO L., "An examination of semantic relationships between professionally assigned metadata and user-generated tags for popular literature in complementary and alternative medicine", *Information Research*, vol. 16, no. 3, 2011.

Text 29 – IYER H., GUADRÓN M., "Older adults and eHealth literacy: Challenges to knowledge representation", in BABIK W. (ed.), *Advances in Knowledge Organization, vol. 14 (2014) – Knowledge Organization in the 21st Century: Between Historical Patterns and Future Prospects – Proceedings of the Thirteenth International ISKO Conference 19-22 May 2014 Kraków, Poland*, Ergon, Würzburg, 2014.

Text 30 – IYER H., "Alternative system of medicine, Ayurveda: Challenges to knowledge organization and representation", in CHAVES GUIMARÃES J.A., OLIVIERA MILANI S., DODEBEI V. (eds), *Advances in Knowledge Organization, Vol. 15 (2016) – Knowledge Organization for a Sustainable World: Challenges and Perspectives for Cultural, Scientific, and Technological Sharing in a Connected Society – Proceedings of the Fourteenth International ISKO Conference 27-29 September 2016 Rio de Janeiro, Brazil*, Ergon, Würzburg, 2016.

Text 31 – IYER H., RAGHAVAN K.S., "Medical ontology: Siddha system of medicine", in RIBEIRO F., CERVEIRA M.E. (eds), *Advances in Knowledge Organization, Vol. 16 (2018) – Challenges and Opportunities for Knowledge Organization in the Digital Age – Proceedings of the Fifteenth International ISKO Conference 9-11 July 2018 Porto, Portugal*, Ergon, Baden-Baden, 2018.

Text 32 – JORENTE M.J.V., "Information Design as knowledge technology in the organization of digital information environments on the Zika Virus and its effects", in RIBEIRO F., CERVEIRA M.E. (eds), *Advances in Knowledge Organization, Vol. 16 (2018) – Challenges and Opportunities for Knowledge Organization in the Digital Age – Proceedings of the Fifteenth International ISKO Conference 9-11 July 2018 Porto, Portugal*, Ergon, Baden-Baden, 2018.

Text 33 – KHOO C.S.G., NA J.-C., WANG V.W., CHAN S., "Developing an ontology for encoding disease treatment information in medical abstracts", *DESIDOC Journal of Library & Information Technology*, vol. 31, no. 31, pp. 103-115, 2011.

Text 34 – KIPP M.E.I., "Tagging of biomedical articles on CiteULike: A comparison of user, author and professional indexing", *Knowledge Organization*, vol. 38, no. 3, pp. 245-261, 2011.

Text 35 – LEE D.H., SCHLEYER T., "Social tagging is no substitute for controlled indexing: A comparison of Medical Subject Headings and CiteULike tags assigned to 231,388 papers", *Journal of the American Society for Information Science and Technology*, vol. 63, no. 9, pp. 1747-1757, 2012.

Text 36 – LIU Y.-H., WACHOLDER N., "Evaluating the impact of MeSH (Medical Subject Headings) terms on different types of searchers", *Information Processing & Management*, vol. 53, no. 4, pp. 851-870, 2017.

Text 37 – LÓPEZ-MERTZ E.M., "The adequacy of the structure of the National Library of Medicine Classification Scheme for organizing pharmacy literature", *Library Resources & Technical Services*, vol. 41, no. 2, pp. 123-135, 1997.

Text 38 – MARTELETO R.M., DOS SANTOS CARVALHO L., "Health as a knowledge domain and social field: Dialogues with Birger Hjørland and Pierre Bourdieu", *Knowledge Organization*, vol. 42, no. 8, pp. 581-590, 2015.

Text 39 – MARCONDES C.H., DA COSTA L.C., "A model to represent and process scientific knowledge in biomedical articles with semantic web technologies", *Knowledge Organization*, vol. 43, no. 2, pp. 86-101 2016.

Text 40 – MCILWAINE I.C., WILLIAMSON N.J., "Medicine and the UDC: The process of restructuring", in ARSENAULT C., TENNIS J.T. (eds), *Advances in Knowledge Organization, Vol. 11 (2008) – Culture and Identity in Knowledge Organization – Proceedings of the Tenth International ISKO Conference 5-8 August 2008 Montréal, Canada*, Ergon, Würzburg, 2008.

Text 41 – MCTAVISH J.R., FORTIER A., "A domain-analytic perspective on sexual health in LCSH and RVM", in BEGHTOL C., SMIRAGLIA R. (eds), *Proceedings from North American Symposium on Knowledge Organization*, vol. 3, pp. 83-93, Toronto, Canada, June 16-17, 2011.

Text 42 – MCTAVISH J.R., NEAL D.R., WATHEN C.N., "Is what you see what you get? Medical Subject Headings and their organizing work in the violence against women research literature", *Knowledge Organization*, vol. 38, no. 5, pp. 381-397, 2011.

Text 43 – MU X., LU K., RYU H., "Explicitly integrating MeSH thesaurus help into health information retrieval systems: An empirical user study", *Information Processing & Management*, vol. 50, no. 1, pp. 24-40, 2014.

Text 44 – MURESAN S., KLAVANS J.L., "Inducing terminologies from text: A case study for the consumer health domain", *Journal of the American Society for Information Science and Technology*, vol. 64, no. 4, pp. 727-744, 2013.

Text 45 – NA J.-H., LENG NEOH H., "Effectiveness of UMLS semantic network as a seed ontology for building a medical domain ontology", *Aslib Proceedings*, vol. 60, no. 1, pp. 32-46, 2008.

Text 46 – NICOLE-GENTY G., "Les classifications médicales", *Bulletin des bibliothèques de France*, vol. 13, no. 7, pp. 289-313, 1968.

Text 47 – OH K., JOO S., JEONG E.-J., "Online consumer health information organization: Users' perspectives on faceted navigation", *Knowledge Organization*, vol. 42, no. 3, pp. 176-186, 2015.

Text 48 – PINTO V., RABELO C., GIRÃO I., "SNOMED-CT as standard language for organization and representation of the information in patient records", *Knowledge Organization*, vol. 41, no. 4, pp. 311-318, 2014.

Text 49 – RAGHAVAN K.S., SAJANA C., "NeurOn. Modeling ontology for neurosurgery", in GNOLI C., MAZZOCCHI F. (eds), *Advances in Knowledge Organization, vol. 12 (2010) – Paradigms and Conceptual systems in Knowledge Organization – Proceedings of the Eleventh International ISKO Conference 23-26 February 2010 Rome, Italy*, Ergon, Würzburg, 2010.

Text 50 – RAJ P., "Depth colon classification schedule for anaesthesiology", *International Classification*, vol. 14, no. 2, pp. 77-83, 1987.

Text 51 – REDALEN A., MILLER N., "Evaluating website modifications at the National Library of Medicine through search log analysis", *D-Lib Magazine*, vol. 6, no. 1, 2000.

Text 52 – RIBEIRO C.J.S., DA SILVA D.R., "Knowledge Organization in the health field: An ontology project to improve the information retrieval process", in RIBEIRO F., CERVEIRA M.E. (eds), *Advances in Knowledge Organization, Vol. 16 (2018) – Challenges and Opportunities for Knowledge Organization in the Digital Age – Proceedings of the Fifteenth International ISKO Conference 9-11 July 2018 Porto, Portugal*, Ergon, Baden-Baden, 2018.

Text 53 – SCHOPEN M., "Cross file searching of biomedical databases at DIMDI", in FUGMANN R. (ed.), *Advances in Knowledge Organization, Vol. 1 (1990) – Tools for Knowledge Organization and the Human Interface – Proceedings 1st International ISKO-Conference Darmstadt, Germany, 14-17 August 1990*, INDEKS Verlag, Frankfurt/Main, 1990.

Text 54 – SEETHARAMA S., "Compatibility among medical classification systems: A case study in the classification of cardiovascular diseases", *International Classification*, vol. 12, no. 2, pp. 80-86, 1985.

Text 55 – SMITH J.B., DEACON P., "Precision with ease: Refining thesaurus support for quality health information searching on HealthInsite", *The Australian Library Journal*, vol. 58, no. 1, pp. 47-62, 2009.

Text 56 – SOERGEL D., TSE T., SLAUGHTER L., "Helping healthcare consumers understand: An 'interpretative layer' for finding and making sense of medical information", in FIESCHI M., COIERA E., LI Y.-C. (eds), *MEDINFO Proceedings of the 11th World Congress on Medical Informatics*, IOS Press, Amsterdam, 2004.

Text 57 – SOLOMON P., "On the use of research categorizations as the basis for organizing knowledge: A test in the domain of information behavior in health care", in MUSTAFA EL-HADI W., MANIEZ J., POLLITT S.A. (eds), *Advances in Knowledge Organization, Vol. 6 (1998) – Structures and Relations in Knowledge Organization – Proceedings of the Fifth International ISKO Conference 25-29 August 1998 Lille, France*, Ergon, Würzburg, 1998.

Text 58 – SUNDIN O., "Towards an understanding of symbolic aspects of professional information: An analysis of the nursing knowledge domain", *Knowledge Organization*, vol. 30, no. 3/4, pp. 170-181, 2003.

Text 59 – SWANSON D.R., SMALHEISER N.R., TORVIK V.I, "Ranking indirect connections in literature-based discovery: The role of medical subject headings", *Journal of the American Society for Information Science and Technology*, vol. 57, no. 11, pp. 1427-1439, 2006.

Text 60 – TANG X., WEI W., LIU G., ZHU J., "An inference model of medical insurance fraud detection: Based on ontology and SWRL", *Knowledge Organization*, vol. 44, no. 2, pp. 84-96, 2017.

Text 61 – TANTI M., ROUX P., CARRIERI M.P., SPIRE B., "Exploiting the knowledge organization of Health 2.0 to create strategic value in public health – An example of application to the problem of drug consumption rooms in France", in CHAVES GUIMARÃES J.A., OLIVIERA MILANI S., DODEBEI V. (eds), *Advances in Knowledge Organization, Vol. 15 (2016) – Knowledge Organization for a Sustainable World: Challenges and Perspectives for Cultural, Scientific, and Technological Sharing in a Connected Society – Proceedings of the Fourteenth International ISKO Conference 27-29 September 2016 Rio de Janeiro, Brazil*, Ergon, Würzburg, 2016.

Text 62 – TANTI M., "The French military documentary system to anticipate health risk: Content and information classification", in RIBEIRO F., CERVEIRA M.E. (eds), *Advances in Knowledge Organization, Vol. 16 (2018) – Challenges and Opportunities for Knowledge Organization in the Digital Age – Proceedings of the Fifteenth International ISKO Conference 9-11 July 2018 Porto, Portugal*, Ergon, Baden-Baden, 2018.

Text 63 – VAKKARI P., "How specific thesauri and a general thesaurus cover lay persons' vocabularies concerning health, nutrition and social services", in GNOLI C., MAZZOCCHI F. (eds), *Advances in Knowledge Organization, Vol. 12 (2010) – Paradigms and Conceptual Systems in Knowledge Organization – Proceedings of the Eleventh International ISKO Conference 23-26 February 2010 Rome, Italy*, Ergon, Würzburg, 2010.

Text 64 – WILLIAMSON N.J., "Complementary and alternative medicine: Its place in the reorganized medical sciences in the Universal Decimal Classification", in MCILWAINE I.C. (eds), *Advances in Knowledge Organization, Vol. 9 (2004) – Global Organization and the Global Society – Proceedings of the Eighth International ISKO Conference 13-16 July 2004 London, UK*, Ergon, Würzburg, 2004.

Text 65 – WOMACK K.R., "Conformity for conformity's sake? The choice of a classification system and a Subject Heading system in academic health sciences libraries", *Cataloging & Classification Quarterly*, vol. 42, no. 1, pp. 93-115, 2006.

Text 66 – WOODS J.W., SNEIDERMAN C.A., HAMEED K., ACKERMAN M.J., HATTON C., "Using UMLS metathesaurus concepts to describe medical images: Dermatology vocabulary", *Computer in Biology and Medicine*, vol. 36, no. 1, pp. 89-100, 2006.

Text 67 – YITZHAKI M., SHAHAR T., "A draft version of a consolidated thesaurus for the rapidly – Growing field of alternative medicine", *66th IFLA Council and General Conference*, Jerusalem, Israel, 13-18 August 2000.

Text 68 – ZENG M.L., KRONENBERG F., MOLHOLT P., "Toward a conceptual framework for complementary and alternative medicine: Challenges and issues", *Knowledge Organization*, vol. 29, no. 1, pp. 27-40, 2001.

Text 69 – ZHANG J., WOLFRAM D., WANG P., HONG Y., GILLIS R., "Visualization of health-subject analysis based on query term co-occurrences", *Journal of the American Society for Information Science and Technology*, vol. 59, no. 12, pp. 1933-1947, 2008.

Text 70 – ZHANG J., WOLFRAM D., "Visual analysis of obesity-related query terms on HealthLink", *Online Information Review*, vol. 33, no. 1, pp. 43-57, 2009.

Text 71 – ZHITOMIRSKY-GEFFET M., EREZ E.S., JUDIT B.-I., "Toward multiviewpoint ontology construction by collaboration of non-experts and crowdsourcing: The case of the effect of diet on health", *Journal of the American Society for Information Science and Technology*, vol. 68, no. 3, pp. 681-694, 2017.

8.12. Appendix 2: MKOS corpus

Bliss Bibliographic Classification, 2nd edition – Bliss Classification Association, "The Bliss Bibliographic Classification: Schedules", available at: https://www.blissclassification.org.uk/bcsched.shtml/, 2016.

Classification Commune des Actes Médicaux – Assurance Maladie, "Classification Commune des Actes Médicaux", available at: https://www.ameli.fr/accueil-de-la-ccam/index.php, 2020.

Colon Classification – ISKO Italia, "Colon Classification – L Medicine", available at: http://www.iskoi.org/doc/colon.htm#L/, 2006.

Dewey Decimal Classification – OCLC, "Dewey Services – Resources", available at: https://www.oclc.org/en/dewey/resources.html, 2020.

Health Topics – MedlinePlus, "Health Topics", available at: https://medlineplus.gov/healthtopics.html/, 2017.

International Classification of Diseases-9 – CDC – National Center for Health Statistics, "International Classification of Diseases, Ninth Revision (ICD-9)", available at: https://www.cdc.gov/nchs/icd/icd9.htm/, 6 November 2015.

International Classification of Diseases-10 – WHO, "CIM-10 Version: 2008", available at https://icd.who.int/browse10/2008/en, 2008.

International Classification of Diseases-11 – WHO, "ICD-11 for Mortality and Morbidity Statistics (Version 04 / 2019)", available at: https://icd.who.int/browse11/l-m/en/, 2019.

Library of Congress Subject Headings – Library of Congress, "Linked Data Service – Library of Congress Subject Headings", available at: http://id.loc.gov/authorities/ subjects.html, 2019.

MeSH Bilingue – INSERM, "Le MeSH bilingue anglais-français", available at: http://mesh.inserm.fr/FrenchMesh/, 2019.

NLM Classification – National Library of Medicine, "NLM Classification 2020 Winter Edition", available at: https://www.nlm.nih.gov/class/index.html/, 2020.

SNOMED – Snomed International, "Leading healthcare terminology, worldwide", available at: http://www.snomed.org/, 2020.

Universal Decimal Classification – UDC Consortium, "UDC Summary", available at: http://www.udcsummary.info/php/index.php, 2020.

List of Authors

Chérifa BOUKACEM-ZEGHMOURI
ELICO
University of Lyon
France

Aude CHAUVIAT
GRESEC
Grenoble Alpes University
France

Viviane CLAVIER
GRESEC
Grenoble Alpes University
France

Hans DILLAERTS
LERASS
Paul Valéry University
Montpellier 3
France

Stefan EK
Åbo Akademi University
Finland

Kristina ERIKSSON-BACKA
Åbo Akademi University Library
Finland

Pascal FOUR
Doctor in Ophthalmology
Romans-sur-Isère
France

Claudio GNOLI
Library of Science and Technology
University of Pavia
Italy

Amar LAKEL
MICA
Bordeaux Montaigne University
Pessac
France

Olivier LE DEUFF
MICA
Bordeaux Montaigne University
Pessac
France

Cécile LORIATO
CEDITEC
Paris-Est Créteil University
France

Céline PAGANELLI
LERASS
Paul Valéry University
Montpellier 3
France

Marcin TRZMIELEWSKI
LERASS
Paul Valéry University
Montpellier 3
France

Eloria VIGOUROUX-ZUGASTI
MICA
Bordeaux Montaigne University
Pessac
France

Index

D, G, H

I, J

K, M, O

P, S

Other titles from

iSTE

in

Health Engineering and Society

2021

BÉRANGER Jérôme
Societal Responsibility of Artificial Intelligence: Towards an Ethical and Eco-responsible AI
(Technological Prospects and Social Applications Set – Volume 4)

CORDELIER Benoit, GALIBERT Olivier
Digital Health Communications
(Technological Prospects and Social Applications Set – Volume 5)

NADOT Michel
Discipline of Nursing: Three-time Knowledge

2020

BARBET Jacques, FOUCQUIER Adrien, THOMAS Yves
Therapeutic Progress in Oncology: Towards a Revolution in Cancer Therapy?
(Health and Patients Set – Volume 2)

MORILLON Laurent, CLAVIER Viviane
Health Research Practices in a Digital Context
(Health Information Set – Volume 4)

PAILLIART Isabelle
New Territories in Health
(Health Information Set – Volume 3)

2019

CLAVIER Viviane, DE OLIVEIRA Jean-Philippe
Food and Health: Actor Strategies in Information and Communication
(Health Information Set – Volume 2)

PIZON Frank
Health Education and Prevention
(Health and Patients Set – Volume 1)

2018

HUARD Pierre
The Management of Chronic Diseases: Organizational Innovation and Efficiency

PAGANELLI Céline
Confidence and Legitimacy in Health Information and Communication
(Health Information Set – Volume 1)

2017

PICARD Robert
Co-design in Living Labs for Healthcare and Independent Living: Concepts, Methods and Tools

2015

BÉRANGER Jérôme
Medical Information Systems Ethics

Printed and bound by CPI Group (UK) Ltd, Croydon, CR0 4YY

13/09/2026

14961036-0004